REBUILDING AMERICA

AMERICA
Volume 1

REBUILDING AMERICA

Volume 1

Planning and Managing Public Works in the 1980s

Roger J. Vaughan and Robert Pollard

BARBARA DYER, GENERAL EDITOR
NORMA DEFREITAS, PRODUCTION MANAGER

Volumes One and Two

THE COUNCIL OF STATE PLANNING AGENCIES
HALL OF THE STATES
400 NORTH CAPITOL STREET
WASHINGTON DC 20001

363
V 36 n
v. 1

On the cover: *Way to the Citadel* by Paul Klee. Courtesy of the Phillips Collection, Washington, D.C.

© 1984 by the Council of State Planning Agencies
Library of Congress Cataloging in Publication Data

Vaughan, Roger J.
Planning and managing public works in the 1980s.

(Rebuilding America; v. 1)
Bibliography: p.
Includes index.
1. United States—Public Works—Management.
I. Pollard, Robert, 1945–   . II. Title.   III. Series:
Vaughan, Roger J. Rebuilding America; v. 1.
HD3885.V36 1983 VOL. 1   363'.0973s   84–1892
[HD3887]   [350.86'0973]
ISBN 0-934842-22-1

Book and Cover Design by Elizabeth Elliott
Typography by The TypeWorks

Manufactured in the United States

The Council of State Planning Agencies is a membership organization comprised of the planning and policy staff of the nation's governors. Through its Washington office, the Council provides assistance to individual states on a wide spectrum of policy matters. The Council also performs policy and technical research on both state and national issues. The Council was formed in 1966; it became affiliated with the National Governors' Association in 1975.

Funding support for this volume was received in part from the Department of Housing and Urban Development under the Financial Management Capacity Sharing Program. The statements, findings, conclusions, recommendations, and other data contained in this report do not necessarily represent the views or policies of the Council of State Planning Agencies. Reproduction of any part of this volume is permitted for any purpose of the United States Government.

The Council of State Planning Agencies
Hall of States
400 North Capitol Street
Washington, D.C. 20001
(201) 624-5386

Foreword

LOCAL GOVERNMENTS face challenging tasks in the 1980s. They have to devise realistic strategies for their cities and mobilize a broad range of resources. They must reexamine the services they provide to establish which are necessary, set priorities for service delivery, determine which services can best be provided by local government directly and which are candidates for alternative service delivery approaches, and actively promote public/private community partnerships.

To help communities address these challenges, the Administration is assisting state and local governments in their efforts to strengthen their management capabilities. HUD's Governmental Capacity Sharing Program is doing this by collecting and disseminating sound ideas for workable practices from the nation's local public and private sectors.

Rebuilding America, Volume I, is part of the Governmental Capacity Sharing Program and is dedicated to encouraging local initiative, self-reliance, and improved performance.

Preface

THE DECAY OF THE NATION'S INFRASTRUCTURE is a major public policy issue. Reports of deteriorating roads, bridges, water supply and treatment systems, ports, and terminals have documented the need to increase spending on maintenance and repair, but by pointing out the dangers of serious environmental degradation and physical harm. Yet, the rate of investment by federal, state, and local government in public capital—both on new projects and on the maintenance of existing facilities—has declined by nearly 50 percent in the last decade.

Larger expenditures for maintenance and repair will not suffice; substantial investments in new facilities are also needed. In some areas, the expansion of employment and population has outstripped the capacity of transportation networks and water treatment plants, while in others, outmigration or the transformation of the economy from an industrial to a service base has resulted in public facilities that are no longer suitable. In addition, new types of public facilities must be built to provide safer and more efficient disposal of industrial and municipal waste, and to accommodate the rapidly growing demands for telecommunications.

A sizable gap separates available state and local fiscal resources and the level of required expenditures. Finances are not the only constraint. Few state and local governments are equipped to assess systematically the condition of their existing infrastructure or to project future demands. Even fewer can effectively evaluate alternative public projects and coordinate capital spending.

This book is intended to serve as a guide to state and local policymakers seeking to rebuild existing public works and to provide new facilities necessary for a changing economic environment. Two distinct areas are investigated: (1) how to raise funds efficiently and equitably, and (2) how to spend these funds more effectively. This volume addresses the latter question—the planning for future public facilities and the management of present public facilities. More specifically, it examines how to choose between competing projects, anticipate future capital spending needs, coordinate capital budgeting, and design alternative administrative structures for managing public facilities. The companion volume, *Financing Public Works in the 1980s*, examines ways of financing public works investments, including initiatives to cut debt financing costs, user fees, techniques for sharing costs with private developers, and the use of lease arrangements and service contracts.

There is no panacea to the infrastructure crisis. More efficient planning, capital budgeting, and facility management practices will have to be adopted; user fees will have to be imposed or increased; new financing mechanisms will have to be designed; responsibility will have to be shared more equally with the private sector; some taxes will have to be raised. More radical approaches must also be sought, including reconsideration of engineering, maintenance, repair and replacement standards; less use of tax-exempt bond revenues to subsidize private development; efficient pricing of public facilities and services; reduced use of tax abatements and incentives; and increased sharing of project costs with private developers.

A carefully considered public investment strategy that draws upon all these elements can resolve the crisis, arrest the rate of deterioration in the nation's public works, and place public capital investments on an efficient, businesslike basis.

Acknowledgments

THIS BOOK WAS PREPARED under a contract from the U.S. Department of Housing and Urban Development (HUD). Alan Siegel and Ed Stromberg in the HUD Office of Policy Development and Research provided comments and suggestions throughout this study. The book draws heavily on material prepared under two other HUD-sponsored research projects conducted by the Urban Institute and by the National Conference of State Legislatures (NCSL). The authors are indebted to George Peterson of the Urban Institute and to Ken Kirkland of NCSL for sharing their research.

Many others have assisted by reviewing drafts including Barbara Dyer, Council of State Planning Agencies, Hugh O'Neill, Deputy Secretary to the Governor of New York, Mark Ferber, Kidder Peabody and Co., Bernard Johnson, Assistant Director of Planning, State of Vermont, Susan Tixier, Office of the Governor, New Mexico, Dr. Richard Mudge, Principal Analyst, Congressional Budget Office, and Walt Plosila, Deputy Secretary, Department of Commerce, Pennsylvania.

Many other state officials assisted the project by providing information on state practices. Finally, Norma de-Freitas of the Council of State Planning Agencies assisted at all stages of the project and in the preparation of the book.

In spite of this assistance, the authors remain responsible for remaining errors and omissions and for the views expressed, which are not necessarily those of the Department of Housing and Urban Development or the Council of State Planning Agencies.

Contents

List of Tables and Figures

REBUILDING AMERICA

CHAPTER I

The Infrastructure "Crisis"

IN 1982, the condition of the nation's public capital became front page news. *Newsweek* (8/2/82) ran a cover story on "The Decaying of America"; *U.S. News and World Report* labelled public works a $2.5 trillion problem; *The Wall Street Journal* featured stories on the decay of New York City's bridges (8/11/82) and on the 10,000 school children who must walk across bridges because school buses cannot carry them across safely (10/1/82); and *Harpers* called for the revival of President Roosevelt's Works Progress Administration to help rebuild America (October, 1982). Study after study has documented the need for massive expenditures to arrest the rate of deterioration of highways, roads, bridges, waste and sewage systems, dams, ports, and terminals.[1] The collapse of the bridge over the Mianus River in Connecticut has underlined the threat to safety of the neglect of public works.

The term infrastructure includes a wide array of public facilities and equipment required to provide social services and support private sector economic activity. Among the types of public service and production facilities commonly included in infrastructure are roads, bridges, water and sewer systems, airports, ports and public buildings (see Table 1). A jurisdiction, however, may choose a broader or narrower definition for capital facilities planning and management activities, depending on what is practical, feasible, or necessary.

Much of the responsibility for rebuilding America will fall on state and local governments. Federal grants have recently been curbed, and with growing federal deficits, the prospects for a significant increase in aid from Washington are dim. The recent gasoline tax legislation is a

1

Table 1

Categories of Public Infrastructure

Service Facilities	Production Facilities

Service Facilities

Education
- Elementary Schools
- Middle Schools
- Secondary Schools
- Public Libraries

Health
- Hospitals
- Nursing Home
- Ambulatory (Outpatient) Care Facilities
- Ambulatory Dental Care Facilities
- Ambulatory Mental Health Facilities
- Residential Facilities for:
  - Orphans and dependent children
  - the emotionally disturbed
  - alcoholics and drug abusers
  - the physically handicapped
  - mentally retarded
  - blind and deaf
- Emergency Vehicle Service

Justice
- Law Enforcement Facilities
- Jails

Recreation
- Community Recreation Facilities

Transportation
- Railroad Facilities
- Airport and Related Facilities
- Streets and Highways (including bridges)
- Inter- and Intra-Community Transit

Production Facilities

Energy
- Direct Power Suppliers

Fire Safety
- Fire Stations
- Vehicles
- Communications System
- Water Supply and Storage

Solid Waste
- Collection Facilities and Equipment
- Disposal Sites

Telecommunications
- Cable Television
- Over-The-Air Television
- Disaster Preparedness

Waste Water
- Sewer Mains and Collection Systems
- Treatment and Disposal Systems

Water Supply
- Community Systems
  - Storage Facilities
  - Treatment Facilities
  - Delivery Facilities
- On-Site Wells and Cisterns

Source: Abt Associates, *National Rural Community Facilities Assesment Study: Pilot Phase, Final Report*, March, 1980, p. 16.

notable but limited exception. Yet state and local governments face severe budget problems of their own. Prolonged recession has eroded tax revenues. Widespread federal cuts have delegated to state capitols and city halls much broader fiscal and administrative responsibilities for public services and facilities. And high real interest rates have made borrowing prohibitive for all but the most urgent projects.

Most of the public debate has revolved around the staggering cost of arresting the rate of deterioration of the nation's public works. But even if more money were somehow available, there is no assurance it would be spent wisely. The piecemeal approach to public investments has led to waste and inefficiency in the following ways:

• Coordination of spending plans is difficult when responsibility for public works investments is shared among 100 federal agencies, hundreds of state agencies, 3,042 county governments, 35,000 general-purpose local governments, 15,000 school districts, 26,000 special districts, and 209 multi-state organizations. Few of these entities have the basic processes and procedures for anticipating the future, planning capital investments, and managing public facilities.[2]
• Elected officials have short time horizons and are thereby tempted to defer spending that has a distant payoff. Money spent on maintaining a sewer system is "invisible"—at least until the sewer ruptures. Money spent on teachers and police is highly visible.
• Federal policies for financing public works discourages proper operation and maintenance. Federal highway grants pay 90 percent of the costs for the construction of interstate highways, 75 percent for primary, secondary, and urban road construction and 80 percent for bridge replacement and major rehabilitation. Wastewater treatment grants provide 75 percent of the construction costs of eligible projects (falling to 55 percent by 1985). Neither of these programs allows federal funds to be used for operation and maintenance, although the highway program does permit the use of federal funds for interstate and state highway resurfacing and rehabilitation. More than 90 percent of the responses to a survey

of state and local governments conducted by the American Public Works Association indicated that federal capital funds tend to cause them to lower their own priorities for maintenance and rehabilitation (U.S. GAO, November 1982).

While better planning, managing and coordinating can stretch public dollars, the gap between current spending levels and the hundreds of billions of dollars that have been called for cannot be spanned. A broader public investment strategy is required—a strategy encompassing the following eight basic elements.

1. *Redesign Engineering Standards and Improve Maintenance Procedures.* Many of the standards that are used by federal, state and local governments to measure investment needs and to guide maintenance procedures are based upon engineering criteria, not upon economic criteria. This makes it difficult to assign criteria to the long list of investment needs. These standards should be modified to allow money to be spent on high priority projects. In an analysis of public investment financing policy, Peterson and Miller conclude (1981, p. 2):

> Although needs gaps are usually thought to be filled by additional spending, they can be closed by reconsidering and reducing the needs standards that have given rise to the investment gap. This is not a matter of redefining infrastructure needs to make them disappear, but of recognizing that needs always exceed resources and that, with budgets as tight as they now are, priorities must be selected even within "needs" categories.

The issue is not only defining replacement cycles that reflect realistic priorities, but also establishing standards that are cost-effective. For example, Rochelle Stanfield (1982, p. 2017) reported that:

> The Washington Suburban Sanitary Commission which supplies water and sewage treatment for two counties in Maryland, discovered, for example, that by lowering from 95 percent to 92 percent the chance that there would be a sufficient water supply in a severe drought, the cost to

maintain the necessary reserve of water could be slashed from $440 million to $31 million.

2. *Improve Capital Planning, Budgeting and Managing.* A comprehensive public investment strategy will require sophisticated planning and budgeting procedures and the design of effective management techniques for disbursing funds and for operating public facilities. Shepard and Goddard (no date) summarize the need for better planning:

> Strategies and programs to restore the infrastructure to its role of providing efficient service to human and economic goals must be preceded by deliberation and the capacity to act—the country has learned all too well that massive doses of money, even if they were available, can contain the disease but cannot provide a complete cure. Time conscious planning and the capacity to analyze appropriate capital development strategies can be as valuable a step as the generation of the dollar resources needed for restoration and revitalization.

The failure to develop comprehensive needs assessment, planning and budgeting techniques results not only in the misallocation of resources among projects and the failure to prevent serious breakdowns, but also in increasing delays in undertaking projects. In *As Time Goes By,* Dr. Pat Choate (1979) documents a backlog of funded but unconstructed federal, state, and local capital projects of $80-$100 billion. Escalating real costs of construction and foregone earnings on idle funds increases project costs. Effective planning mechanisms for capital projects need to be developed as well as mechanisms to speed up the permitting and contract-letting procedures to avoid delays. New administrative structures are needed to ensure that existing public facilities are efficiently managed.

3. *Reduce Subsidies to Private Investment.* State and local governments must cut back on the use of tax revenues and the proceeds from tax-exempt bond sales to subsidize private sector capital investments. In 1981, more than half the proceeds from long-term bond sales were used to finance projects that traditionally are not the responsibility of the public sector, including private hospitals, industrial development projects, pollution control facilities for private companies, convention centers and university dormitories (see Vol. 2, Financing, Chapter 5). If state and local governments are to set aside the resources necessary to build and maintain traditionally public infrastructure—roads, bridges, education and health facilities, and water supply and treatment facilities—they must cut back on these subsidies. They must also look at opportunities for turning over some of the responsibility for developing and maintaining certain public facilities to private firms.

4. *Provide Project-Specific Cost-Sharing Arrangements with Private Firms.* Some public capital projects are made for the exclusive benefit of major private sector developments. Obvious examples are new towns constructed to house the labor force for large-scale energy developments in the West. Smaller projects include streets, sidewalks and utility hook-ups for residential subdivisions, and highway and railroad spurs for new manufacturing facilities. The beneficiaries of these investments should be required to pay at least part of the investment costs.

5. *Charge for Public Services.* Bridging the infrastructure gap will require greater reliance on user fees for the use of public facilities such as waste disposal, water supply, and recreation facilities. User fees not only produce revenues, but also promote a more efficient use of public facilities. Only when consumers pay for each gallon of water used, are they more careful in how much they consume. The resulting conservation may reduce the need to construct expensive new capacity.

6. *Improve Bond Financing Mechanisms.* A little more than half of all state and local capital spending is financed by the sale of tax-exempt general obligation and limited obligation bonds (see Vol. 2, Financing, Chapter 3). Lack of coordination among borrowers—there are over 1.5 million different tax-exempt issues outstanding—increases financing costs. The costs of issuing tax-exempt bonds can be reduced through state bond banks, state bond guarantees, and state loans to localities. In addition, new revenue sources can be found to service the debt and improve the credit rating.

7. *Increase State Assistance to Local Governments.* Local governments—cities, counties, and special districts—directly invest about twice as much as state governments in capital projects (although funds are often provided from federal and state sources). Many local jurisdictions have reached statutory or constitutional bonding limits, and face extreme fiscal pressures as a result of the economic recession and cuts in federal aid. Some increase in state aid (both in dollars and in planning and technical assistance) will be necessary for those cities and counties with extensive infrastructure needs and slender fiscal resources. The state can help through technical and planning assistance, intrastate revenue sharing, direct loans, categorical project grants, or state assumption of local responsibilities.

8. *Increase Public Capital Investment.* The final element in bridging the gap is increased spending on public works. The revenues will come from user-fees, increases in existing tax rates, and perhaps from reducing tax abatement and exemption programs. Although these steps will not be politically popular, taxpayer opposition may be reduced if the tax increases are directly committed to specific purposes and projects.

These eight elements constitute an effective strategy for planning, financing, and managing public works investments. Developing new planning, budgeting, and managing policies to deal effectively with an issue as broad and as large as public capital investments will require far-reaching changes in present policies and practices. This book and the companion volume, *Financing Public Works in the 1980s*, are intended to help state and local officials charged with responsibility for managing, planning, budgeting, and developing policy to think strategically about public capital investment policies. This volume analyzes the first three elements in the list presented above. It discusses how demands for public services and facilities are changing and reviews ways to set up capital planning and capital budgeting procedures. It also assesses the advantages of alternative administrative structures that can be used to manage public facilities and examines how public works can be timed to provide additional employment during recessions. The companion volume describes alternative policies that can be pursued to impose user fees for public facilities, to improve the operation of the bond market, to share financing responsibilities with the private sector, and to increase state assistance to local governments (elements 4, 5, 6, 7 and 8).

A rational approach to planning, budgeting, and managing requires policymakers to anticipate emerging demands for public facilities. The subsequent chapters of this volume provide insight into these demands. Chapter 2 summarizes the factors that are propelling economic development and describes how these changes may shape future demands for public infrastructure. To translate these broad changes into a capital plan, policymakers must be able to analyze the cost and benefits of specific investment projects. Chapter 3 discusses the political framework within which planning must be conducted. It analyzes what can be expected of a long-term plan and who should be engaged in planning activities. Chapter 4 outlines the economic approach to public decisionmaking, it describes how economic principles can be applied to public works investments, and explores the issues involved in evaluating projects to determine whether they are prudent investments of public funds. The rigorous application of cost-benefit analysis allows planners to move beyond the simplistic needs approach and to develop cost-effective criteria for public works investment.

Chapter 5 examines how the fiscal and administrative responsibility for services and facilities can be equitably and efficiently shared between the public and private sectors. "Privatization" has become a controversial issue. Although the manner in which costs and management are shared will vary among states according to custom, constitution, and need, the chapter offers some general principles that can help fiscally pressed states draw upon the resources of private firms for assistance in delivering public services and financing facilities.

Chapter 6 describes how capital planning can be linked to the annual budget process through a capital budget. Again, although much emphasis has been placed on intricate modeling, common-sense rather than computer models is the basis of a sound capital budget. Effective capital budgeting requires four steps: (1) a long-term plan, (2) a mechanism for coordinating capital plans among agencies and authorities, (3) regular assessments of the condition and rate of depreciation of capital facilities, and (4) the separation of capital and operating expenditures.

Chapter 7 focuses on project management issues. It discusses the advantages and disadvantages of different administrative structures—from state infrastructure banks to local special assessment districts. The chapter points out that different types of agencies behave very differently with respect to the operation and maintenance of public facilities. For example, independent authorities are less likely to neglect maintenance than are state agencies, but their activities are more difficult to coordinate within a state's overall plans and may be less accountable. The appropriate choice of administrative structure will depend upon the overall structure of state government and upon the type of revenue sources that can be earmarked for debt service and operation and maintenance.

Chapter 8 discusses how states may use public works countercyclically. It analyzes why federal programs to fight recessions have failed and how states, by setting up stabilization funds, can dampen local cycles in construction activity. The final chapter summarizes the major implications of the preceding chapters.

The theme that unites the discussions presented here is that state governments must move away from a piecemeal approach to making public capital decisions and move toward an integrated public investment strategy. While this cannot guarantee that no bridges collapse or that no sewer pipes rupture, it will ensure that scarce public resources are used more effectively in meeting the nation's public infrastructure needs.

CHAPTER I NOTES

1. The annotated bibliography provides a list of these studies.

2. For a complete list of the number of agencies engaged in public capital investments see the *1982 Census of Governments*, U.S. Bureau of the Census, Washington, D.C.

CHAPTER II

Anticipating the Future

A CAPITAL INVESTMENT STRATEGY must look beyond replacing existing structures and systems and respond to changing demands for public facilities and services. Future demands change in response to shifts in the structure of the economy, demographic changes, and changes in the preferences of households and businesses. Anticipating future developments and tracing their possible consequences on the demand for public facilities is an integral part of the planning process. The process of capital planning and budgeting requires public officials to make decisions within a much longer time horizon:

> By its very nature, the systematic review and scheduling of capital projects forces decision makers into a long-term consideration of the economic and demographic trends, land-use patterns, and future revenues which a state can anticipate (Petersen, 1977, p. 17).

The past decade has provided painful lessons about the costs of failing to anticipate change. The "urban crisis" resulted, in part, from the failure of large, older cities and industrial states to recognize the strength of the forces promoting suburbanization and regional decentralization. This failure left them overcommitted to existing public services and unable to adapt to the underlying changes occurring in their employment bases. Although public policies could not have reversed the movement, better planning could have reduced the devastating fiscal consequences. Other areas failed to anticipate the consequences of rapid growth, leaving them with clogged traffic arteries, a deteriorating environment, and threatened water supplies. This chapter explores the likely impacts of future changes on the demand for public capital facilities. The U.S. economy is undergoing rapid and dramatic structural changes that will continue throughout the next decade. These changes will lead to significant shifts in the demands for different types of public capital investments. Planning cannot guarantee that the correct decision will always be made. Mistakes inevitably occur in a rapidly changing environment. Careful planning can, however, make more effective use of information and provide projections by which many unnecessary and costly errors can be avoided. Planning is especially critical during periods of rapid economic change when traditional practices are likely to produce unreliable results. The first four sections of this chapter discuss the following four major factors shaping the paths of local and national economic development:

- *Technological Advances.* Coincident advances in data processing and communications technology are dramatically reducing the costs of storing, processing, and transmitting information. The development of new materials will lead to the replacement of structural steel by more durable compounds using silicon and carbon. Biotechnology promises significant breakthroughs in medicine and agriculture.

- *Demographic Changes.* The population and labor force will grow much more slowly in the 1980s than in the 1970s. By 1985, labor force growth will approach zero, falling from a 2 percent rate of growth in 1975. The rate of growth of households will exhibit a similarly sharp decline. The demand for residential infrastructure will decline.

- *World Trade.* The growing importance of world trade will create rapid declines in employment in "import substitute" goods and services and rapid expansion in exports. Severe local economic and fiscal dislocation may result.

- *Resource Management.* Resource development and management in the 1970s were dominated by rapid increases in oil prices; conservation and the development of alternative energy sources mean that sharp real increases in oil prices are unlikely during the next decade.

11

Maintaining arable land and topsoil, ensuring adequate water supplies, and the safe disposal of hazardous waste are becoming significant resource management issues that will require federal, state, and local government action.

These changes will alter the structure of state economies and the demand for public facilities. States must develop a long-term planning horizon in order to anticipate accurately what type of public investments must be undertaken and where they will be needed.

The intention of this chapter is not to predict the extent and level of effect of each of these factors, but to outline how each of these changes may be related, qualitatively, to the demand for public capital investments. Anticipating the future does not necessarily require complex models or highly priced consultants. Important insights can be achieved in a variety of ways, from informal scanning of current publications to the convening of local experts. Colorado has prepared a capital plan reaching into the next century by setting up a temporary blue ribbon panel consisting of leaders from the State's business, labor, and academic communities. New York has established a permanent council of business and labor leaders to establish the State's economic priorities. New Jersey has established a state agency—the New Jersey Capital Facilities Planning Commission—to prepare long-run assessments of the State's capital needs. These are all formal steps and should not be taken until a clear planning strategy has been devised. The first basic step for those charged with developing a planning strategy is to analyze the experience of other, comparable states. The Bibliography in this book provides an initial guide for these efforts. There is no right way to anticipate the future. States should adopt an eclectic approach to determining economic priorities and planning strategies depending upon their resources and needs.

Technological Advances[1]

The current technological revolution is changing the U.S. economy as dramatically as, but more rapidly than, the industrial revolution of the nineteenth century. The pace of change has accelerated in the past decade, propelled by the convergent developments in data processing and in telecommunications. During the last ten years the costs of storing and processing data have been reduced by more than 99 percent, and computers have become cheap enough to become standard appliances in many small businesses and even households. Telecommunications development will shape development as radically during the next two decades as the interstate highway system has in the past two. It also offers tremendous opportunities to reduce the cost of delivering public services.

More than 50 percent of all jobs are now classified as involving the processing and communication of information. Jobs in manufacturing are a shrinking fraction of total employment. In 1960, more than a third of the work force was engaged in manufacturing; today it is barely one-fifth. By the turn of the century, it may have fallen to one person in ten. Automation is likely to displace many highly paid but low skilled workers, particularly in the manufacturing sector.

But the computer-communications revolution is only the leading edge of the broader technological revolution. Low-cost data processing is producing engineering breakthroughs in many fields by lowering the costs of product research and development. Many research facilities no longer need invest in huge, expensive mainframe computers. Complex basic research projects that were feasible in only a few major centers a decade ago can now be undertaken by small companies and colleges. Many businesses can benefit from productivity-enhancing innovations that, until recently, were limited to large corporations.

Advances in other scientific fields will also be important. Biotechnology, some predict, will lead to cost-cutting innovations in medical treatment, to rapid increases in agricultural productivity, and to new and inexpensive ways to combat pollution. Ceramics and new materials based on silicon and carbon compounds are forecast to produce low-cost and more durable construction materials by the end of the century.

The changes provoked by these new developments will require quick action by state and local governments. They were ill-prepared to deal with the rapid rate of suburbanization after World War II that was the result of new high-way construction and subsidies to homeownership. If state and local governments are to avoid repeating costly past mistakes, they must begin to examine the implications of new technology for public investments.

As yet, little research on how new technology will shape socioeconomic development has been undertaken. But even in the absence of detailed research some likely changes during the next decades can be predicted. Examples may include the following:

- *Health Care.* Cable systems will allow in-home monitoring and diagnosing of chronically ill patients. Large hospitals and clinics may become less important and will be replaced by smaller, neighborhood service centers. At the same time, new health care technologies made possible through genetic engineering and advanced research will extend life expectancy and contribute to the graying of America.
- *Educational Services.* Many courses previously available only in classrooms will be available to anyone with a TV and a cable connection. The "electronic classroom" will require substantial public investments in telecommunications equipment and services, but reduced investments in traditional education facilities.
- *Business Services.* Retail establishments and small businesses will be able to maintain their books and dramatically reduce inventories through two-way cable connections to accountants and large wholesale and warehouse centers. This will reduce demands for space and therefore improve the competitive advantage of central city locations relative to suburban sites.
- *Retailing.* Households will be able to shop for standardized products such as appliances and groceries through cable systems connected to the computer or television. Orders will be able to be placed at any time of the day or night and deliveries made at a time convenient to the household. Large shopping malls may become obsolete, while specialty stores—whose goods cannot be easily purchased by computer—will grow.
- *Working At Home.* The ability of computers and word processors to be linked via telephone and cable systems

will allow an increasing fraction of the work force to work at home. Typists, programmers, stock analysts, or bankers may not need to commute to their offices during rush hours.

- *Hazardous Waste Treatment.* New methods for handling and treating toxic wastes will allow the solution of many critical environmental pollution problems, but new materials and new technologies may contribute new problems.

The rapid rate of technological change will also change the nature of economic development.

- *Educational Institutions.* Increasingly, economic development will occur around universities and colleges and will be based upon the birth of new enterprises. The large industrial park will be less important than smaller facilities conducive to the "incubation" of new businesses. Education and business development will become increasingly interdependent.

- *Classroom and Curriculum Obsolescence.* Many businesses are already setting up their own in-house education and training programs. Universities and colleges are often unable to train their students in state-of-the-art techniques. The student intern in the laboratory of a major corporation may become more common. Curricula in engineering and the physical sciences will be modified to give credit for new courses carried out not on campus but in businesses.

- *Cooperative and Joint Ventures.* New technologies may require more joint ventures. A small machine tool company cannot afford to purchase a Computer Assisted Design/Computer Assisted Manufacturing (CAD-CAM) system, or to invest in learning how to use them. A cooperative of a dozen such companies in conjunction with technicians from a college or university could share a facility and adapt to using the equipment at a much lower cost. A single building could house several businesses, a shared computer, and a classroom. These types of arrangements may become increasingly frequent.

15

These are only a sample of the developments that may shape demands for public capital investments. Demands for traditional transportation infrastructure may decline. Education facilities must be reconfigured. Investments must be made in telecommunications facilities. With the rapid obsolescence of data-processing equipment, state and local governments should consider leasing or service contracting for computer services rather than purchasing the equipment themselves (see *Financing*, Chapter 7). State governments can also play a critical role in helping local governments adopt new ways of delivering health, education, and social services. Technology assessment must become a central part of state investment planning and management.

Demographic Changes

The postwar baby boom generation has now entered the labor force and has purchased homes. Between 1970 and 1980, the work force in the U.S. grew by nearly 25 percent—more than in any other developed country except Canada. In Japan, the growth in the work force was barely 12 percent over the same period, 4 percent in France, and 0 percent in Britain and Germany. As a result, the U.S. experienced a higher rate of unemployment, particularly among new entrants to the labor force, and a slower rate of growth of real income and labor productivity than other developed nations.

Population growth fell from an annual rate of 1.5 percent at the beginning of the decade to barely 0.7 percent at the end. However, the rate of growth of households—a critical factor in determining the demand for housing and related residential infra structure—continued at two or three times the population growth rate. This was the result of the baby boom generation leaving home, shrinking family size (from 3.2 persons per household in 1970 to 2.7 per household in 1980), and more frequent divorces. The labor force grew faster than the population because the rapid growth in labor force participation by women more than compensated for a decline in labor force participation by men—particularly minorities and older people.

The 1980s will be very different. By the end of the

decade, the labor force will not be growing. Population will be increasing very slowly. The work force will be better educated. Two-thirds of those retiring during the past decade had completed no more than eight years of school, while three-fifths of those entering the work force will have received at least some post-secondary education. Book-learning will be substituting for experience gained on the job. But those entering the work force can expect much more volatile careers than of those retiring. The new work force will have to continually update their skills and, in many instances, change jobs, occupations, and industries to keep pace with evolving technology.

These demographic changes will have several effects. First, real incomes of the relatively well-educated will grow more rapidly in the 1980s than in the 1970s, propelled by the demands of high technology industry. Rising income may result in a more rapid gentrification of center cities as the affluent seek housing that is convenient to employment for a two-professional household, and that offers the social and cultural amenities that often influence location decisions. If gentrification happens, cities may consider public investments that enhance the quality of life—such as recreation facilities, environmental improvements, and cultural centers.

At the other end of the income scale, those with limited education may find themselves falling farther behind. During the last 15 years, the share of income earned by the poorest 40 percent of the population has fallen from 12 percent in 1965 to 9 percent today. Technological progress—with its emphasis on educational attainment—will further disadvantage the poorest and least educated members of the population. The economically disadvantaged will face stronger competition in labor markets from the growing number of immigrants—both legal and illegal—attracted by the promise of employment resulting from the slowing growth of the domestic labor force.

Overall, education facilities will experience declining enrollments so that long term plans will have to develop ways of reducing capacity. At the same time, demands for retraining and remedial education will increase. Planners will therefore have to take a comprehensive view of educational needs so that existing facilities can be adopted from

17

areas where demand is declining to areas where it is expanding.

The aging of the population will require increased investments in housing suitable for elderly households so that, although the aggregate demand for housing will grow slowly, there may be some need to downsize housing as the number of one and two-person elderly households grows and the number of facilities with children present declines. The most important effect of the graying of America will be the increased demands for health service facilities—particularly long-term care institutions. Expenditures for medical services have been forecast to rise from 10 percent of the gross national product (GNP) today to 12 percent by 2000.

Growth of World Trade

In the last three decades, the United States has moved from being a relatively closed economy in which international trade had little influence on the domestic economy to being an open economy in which changing world trade patterns have immediate and, often, dramatic impacts on state and local economies. The rebuilding of European and Asian economies following World War II, the industrialization of less developed countries, and the growing awareness of the importance of natural resources have created an increasingly interdependent world economy.

The number of jobs connected with import and export activity has grown from only 6 percent of the labor force in 1950 to 18 percent today. Major U.S. exports include traditional goods such as agriculture and high technology manufacturing equipment, as well as less traditional exports such as financial and management services. At the same time, many established industries have experienced a rapid domestic decline in the face of growing imports—among the most obvious are automobiles, steel, textile, and apparel.

The growth of world trade has been accompanied by increased specialization among nations, with trading partners increasing their output of those goods and services they can produce at a relatively low cost, and increasing imports of those goods and services in which they do not have a comparative advantage. The comparative advantage of the U.S. is clearly in those activities that are land intensive (agriculture) and that require highly educated and skilled labor (services and advanced technology manufacturing). The growing intensity of international competition is likely to cause even more rapid adjustment in the coming decade.

The opening of the national economy has affected capital markets as well as labor markets. Between 1950 and 1980 investments by U.S. companies abroad rose from $19 billion to $377 billion. Most of these investments were made to develop and to secure sources of raw materials for U.S.-based companies. Investments by foreign companies and governments in the U.S. rose from $3.6 billion to nearly $50 billion.

The expansion of international trade and investments has three important implications for infrastructure investments. First, the capacity of marine facilities and airports to handle freight will have to be expanded and adapted to handle modern transportation technologies. Investments in port facilities must be coupled with increased investments in the distribution system that connects local economies with these ports. After several decades of decline, the use of railroads for bulk-cargoes is likely to increase, reinforced by the rising exports of raw materials and food, and the increased imports of basic manufactured goods. State infrastructure investments can play an important role in helping to integrate their existing rail networks as part of their transportation systems.

Second, the growth in trade will reinforce the need for states to increase education and training programs—a coordinated human capital investment strategy—for those workers whose skills are rendered obsolete by increased imports and technological progress. Devastating plant closings will abate as the national economy recovers, but will continue to be a chronic problem in communities whose economic base is concentrated in import-substitute industries. Training facilities, particularly those developed in cooperation with export industries, will play a central role in easing the transition process.

Third, infrastructure investment strategies by state and local governments must focus on growing, export-based, industries. State and local governments must resist the temptation to use public capital resources to prop up companies adversely affected by international trade.

Natural Resources Management

For most of the past two-hundred years, the United States has not faced the binding resource constraints confronted by other developed nations. Land has been available in abundance. Government policy has been directed at limiting agricultural output rather than at husbanding soil resources. Water has been scarce in only a few regions. Energy has been available at a fraction of its cost in Europe. Waste could be disposed of with little apparent harm in landfills, in flowing rivers, and in the air.

During the last decade, the symptoms of poor resource management have intensified. Soil erosion is a major problem in the Midwest—more topsoil than wheat is shipped down the Mississippi each year. Excessive drilling into natural aquifers is causing land subsidence in Arizona and Texas. Insecticides have endangered drinking water supplies on Long Island. Energy costs have multiplied tenfold in less than a decade. The catalogue of hazardous waste disposal sites that are environmentally unsafe grows annually. Many lakes and rivers can bear no more pollutants, and their marine life is endangered.

Addressing these problems will require not only substantial capital investments by state and local governments, but also a rethinking of the way natural resources are managed. No longer can unlimited and underpriced access be allowed to either surface or artesian water by farmers, mining and manufacturing companies, and residential developers. Recreation and preservation areas will have to be managed to prevent overuse. Users of resources—from boaters and backpackers to builders and farmers—will have to pay their way.

Two of the most significant resource issues that will require substantial infrastructure investments are: (1) water supplies and (2) toxic waste disposal. Neither of these areas lends itself easily to a completely private solution. The ownership of water is complex. Access to water will require careful monitoring and pricing by state and local governments. Toxic waste disposal also has legal and environmental implications that deter purely private sector solutions. Increased state capital investments will be required.

Conclusions

Planning is a process of determining what is to be done in the future. It enables decisionmakers to identify potential problems before they occur and to prepare ways of dealing with those problems. Planning must be forward looking. It is too easy to assume that the future will be a simple continuation of the present. But if planners assume this, they will fail to anticipate major economic and social changes that have important implications for public capital investments. While no one can predict the future with certainty, the task of capital planning can be greatly aided by identifying those factors that influence the demands for public facilities and by developing forecasts of how these factors are likely to change. These forecasts can be based upon formal models or informal estimates—in fact, an eclectic approach may prove the most flexible and useful.

CHAPTER II NOTES

1. Awareness of the profound effect that technological change is having on the national and state economies has precipitated a growing number of studies. The Office of Technology Assessment in Washington, D.C., is currently undertaking a broad review of the impact of technological change on regional development. Many states have established commissions, councils, or task forces to address the issue of high technology development. New magazines such as *Technology*, and *Technology Review*, analyze the economic impacts of new developments. Several recent books including *Global Reach: The Future of High Technology in America*, by James Botkin, Dan Dimancescu, and Ray Stata, (Ballinger, Cambridge, Massachusetts, 1982), and *The New Alchemists* by Dick Hanson (Little Brown, New York, 1981) have examined the new industries spawned by new technologies. In March 1983, *Business Week* featured high technology as a cover story (3/19/83). This section draws upon these diverse sources. Unless otherwise indicated, data are from the *U.S. Statistical Abstract*, 1982, Washington, D.C.

CHAPTER III

The Politics and Process of Capital Planning

EFFECTIVE PLANNING can make much better use of scarce dollars in meeting a state's capital investment needs. Not all planning is effective, however, or is used effectively by decisionmakers. Plans may be based upon inadequate or misleading information. They may be developed using inappropriate forecasting or project evaluation techniques. Plans may be misused or ignored in the preparation of capital budgets or by decisionmakers. There are no easy rules to follow that guarantee good planning. The process must be tailored to fit the institutional, constitutional, and political constraints of the state. While there are no rules, there are some general principles that can help in custom-building a planning process. This chapter explores these principles. The first section discusses the context within which cost-benefit analysis can be applied to capital planning—recognizing the political constraints. The subsequent sections describe how to formalize capital planning procedures within the overall decisionmaking and administrative process—including discussions of what is planning, who should be involved in planning, and the elements of a successful planning mechanism.

Planning can be effective only with the commitment of the governor. Only the governor can enlist the participation of business, labor, and academic leaders to support the work of state planners, ensure that all agencies cooperate, insist that long-term plans are integrated into a capital budget and carry forward the specific elements of the plan before the legislature or the public. In this sense, planning is essentially a gubernatorial responsibility.

Public Decisionmaking and Economic Theory

Planners must draw upon economic techniques in evaluating alternative capital projects required to meet the state's infrastructure needs. The most frequently used approach is cost-benefit analysis, applied to select viable projects. This is a logical and theoretically simple approach to public decisionmaking and is described in the following chapter. Cost-benefit analysis has been applied countless times to public as well as private investment decisions and has shaped the way most policymakers think about capital investment decisions. In the real world of policy planning and budgeting, however, this elegant approach is often forgotten or discarded. Some projects must be undertaken with no regard to costs and benefits because of federal or court mandates. Others are financed as a result of political log-rolling, even though they are clearly inefficient undertakings. Conflicts of interagency turf prevent the development of some viable projects. For others, data, staff, and time are often inadequate for more than a superficial analysis of likely outcomes. And finally, as one of the many corollaries of Murphy's Law states, "things do not work out as planned." In short, when planning public capital investments, institutional, constitutional, and political considerations must be weighed together with purely economic factors.

It is easy, however, to assign too much weight to these "noneconomic" factors in planning public capital projects. While major decisions are influenced by the data and analyses prepared by planners, decisions are ultimately made as a result of negotiations between the state's chief executive and the legislature. Information that may or may not be pertinent to a particular decision will be readily available from lobbying and public interest groups. Even before a particular issue becomes public, it may be hotly debated among rival agencies within the administration or even among rival factions within the executive office. If planning and policy analysis are properly conducted, the results should not join the planners to one side or another in the public debate. Planners should concern themselves with providing those engaged in the decisionmaking process with the best possible information, not with committing themselves to one side or another. When planners become lobbyists, they cast doubt on the objectivity of their analyses, and jeopardize their ability to perform their proper function in the decisionmaking process. Good planning can often lead to better decisionmaking. But it is *not* the act of decision-making.

A capital planning document should rarely make a specific recommendation (at least until a project has become part of the governor's capital budget). Instead, it should list the costs, benefits, and other relevant but less quantifiable outcomes of the investment alternatives under consideration. Those that prepared the report may find themselves publicly debating methodology or data, but only rarely are the most intense policy debates concerned with quantifiable outcomes. They are more likely to center on the relative weights attached to aspects of the project for which dollar values cannot easily be assigned— for example, the loss of wilderness land, the shift of employment between two communities, or issues of moral principle. On occasion, further analysis may yield data that can clarify the discussion. As a general rule, capital plans should be based on an analysis that is as rigorous and objective as resources permit.

Perhaps the most difficult dilemma to resolve is the allocation of planning responsibility between the executive branch and the legislative. Both branches will engage in capital planning in preparation for annual or biennial budgets. Yet they may disagree strongly. Building bipartisan support for infrastructure investments is hampered by the tradition of using major construction projects as a political pork barrel. Conflict can be contained if there is some contact between executive and legislative planners and if they maintain a reputation for competence and objectivity. Prolonged conflict will prove harmful to both sides.

Is Capital Planning Useful?

Capital planning is a vague term and is used to refer to a multitude of activities directed at a wide array of goals—from informal predictions of long-term economic and social changes to the preparation of detailed land-use blueprints. In recent years, public sector planning has fallen into disrepute. It has been mistakenly viewed as a substitute for action. In part, this was a reaction to the highly formalized land-use planning that accompanied the massive construction of commuter freeways in urban areas. It was also a result of locating planning activities far outside the decisionmaking process so that plans did not have to be read by those preparing budgets, and planners had little idea what the important issues were confronting the executive branch.

The demise of planning is also due to the mandated planning required by federal programs. In his State of the Union address in February 1981, President Reagan proposed to eliminate the Economic Development Administration (EDA) because it had done little other than support grantsmen, planners, and middlemen. All too often, federal planning requirements, such as EDA's Overall Economic Development Plan and the Employment and Training Administration's Concentrated Employment Plan, for examples, were little more than an extension of the grant application process. That is, instead of using the planning process as a way of defining and evaluating alternative capital investment policies and projects, state and local governments used the mandated plan to justify a project that had already been selected or to satisfy federal requirements for a grant. Ex post planning may provide employment to recent graduates from planning schools, but it does little to improve the decisionmaking process.

While public sector planning is maligned, private corporations have become more and more committed to strategic planning. The purpose of strategic planning is to inform the decisionmaking process by anticipating future changes, by devising possible strategies in advance, and by evaluating the consequences of alternative policies. The rapid and accelerating pace of economic change and a growing sense of the cost of failure to plan have contributed to the increasing importance attached to strategic planning. As the President of Control Data Corporation, told the White House Conference on Strategic Planning:

> This enthusiasm for strategic planning came to business as an evolutionary process rather than a revolutionary

one. It came by the necessity of business having to adapt to changing needs. All this started by having a problem and having to do something different about the problem than was done previously (Walter, ed., 1980, p. 7).

One of the major barriers to setting up a strategic planning operation within state government is the short time horizon of political administration. A private corporation has a longer expected life than the administration of a governor. There will be considerable reluctance to engage in planning that has a time horizon much beyond the next election. In addition, many federal and state capital programs are financed on a year-to-year basis which makes long term planning difficult. However, there are ways around this electoral myopia. One way is to establish a state agency with the responsibility for planning—for example, New Jersey's Capital Facilities Planning Commission or a state infrastructure bank (see Chapter 7). This has a life beyond that of a particular governor. Another approach is to use the resources of public authorities in the planning process (see Chapter 7). These, too, have a long life, and because of their line of responsibilities, they have a longer time horizon than most state agencies. Finally, once a planning process has been established and is working well, it is unlikely to be dissolved by a governor's successor. Once it becomes part of the budget process and is used by all agencies, it will tend to survive.

Strategic planning is potentially useful. It serves two purposes. "First, to develop an integrated, coordinated and consistent long-term plan of action for an organization, and, second, to facilitate the adoption of the organization to environmental change" (Lorange and Vancil, 1979). But most public sector planning activities have not been useful because they have not been properly integrated within the decisionmaking process. Planning will only produce better decisions if the planning process is integrated within the overall budgeting and decisionmaking processes. The following sections suggest ways to design the planning process that makes this integration easier.

What Is Capital Planning?

To those whose understanding of capital planning has been shaped by the unveiling of five-year plans complete with multicolored maps, it is important to understand the following four aspects of strategic planning.

1. *Planning is a continuous process.* It is a way of setting broad objectives, devising appropriate policies, and communicating decisions to those who must respond to them. In an ever changing political, economic and social environment, objectives must be reexamined, policies and programs modified, and the administrative structure revised. Each change must be carried out within a strategic framework. A useful analogy is that of a transatlantic sailing trip. A successful voyage requires more than putting out from harbor, pointing the craft toward the destination and then relaxing with a good book. Constant adjustments in course and rigging are necessary to respond to currents and weather. Contingency arrangements must be made and the crew fully briefed in case of severe weather or structural failure. New procedures may have to be developed according to the abilities of the crew and the characteristics of the boat. The captain who makes these plans and ensures that the crew are aware of the details is not being excessively cautious, but is obeying the basic rules of seamanship. An emergency is not the time to discuss what needs to be done. Even without an emergency, constant monitoring of the boat's position and performance is essential. If the crises that may befall state and local government are not quite as violent and immediate as those that arise aboard the boat, planning is every bit as important and as continuous a function.

2. *The management of the flow of information is a central element of the planning process.* Planning requires two types of knowledge: (1) up-to-date information about prevailing economic and social conditions, and, (2) familiarity with appropriate analytic techniques needed to process this information and evaluate the consequences of alternative decisions. Both types of data must be systematically updated and reviewed. Planning is not forecasting. Forecasting is concerned with determining what the future *will* look like rather than what it *should* look like. The latter is the purpose of planning. Forecasts are only inputs into the planning process.

In addition to the management of information, the planning process requires the effective *communication* of

27

information to several different constituencies, including those who must provide data, those charged with developing alternative policies and projects, those who must act upon the decisions, and the public at large. Planning should be undertaken by a relatively small number of people, a group too small to be aware of the details of the programs and projects relevant to the process. Unless planners clearly communicate the broad objectives of government policy and the issues under consideration, they will not receive all the information they need.

3. *Planning is an alchemist's amalgam of sciences and art* (Amara, 1979). Sophisticated planning and economic models may prove to be useful inputs in the planning process but they do not eliminate the critical role of personal judgment. The benefits of formal models to planners are always limited by the nature of the simplifying assumptions inherent in using models to describe complex reality and by the availability and reliability of the data required to estimate the model. Deciding what model to use and how to apply it to a specific decision is more of an art than a science. As Control Data's Vice President for Planning described it: "By the very necessity of the human component, planning must be pragmatic in that it must adapt to the situation and it must be evolutionary— changing as people change" (Walter, 1980, p. 41).

4. *Setting up a strategic planning process is not always popular because it requires busy managers to do three things that they may find unpleasant: set aside time, think, and write* (Donnelly 1980, p. 4). In a survey of corporate managers, 40 percent stated that "planning was too time consuming, mysterious in method, and useless since the future is uncertain anyhow" (Bologna, 1980, p. 24). But the experience of major corporations is that, if initial planning efforts are difficult, unpopular, and imperfect, the process becomes easier, more popular and more successful over time (see Proceedings of White House Conference on Strategic Planning, 1980). A state's chief executive should not expect the introduction of a planning operation to be greeted with unbounded enthusiasm by agency heads and bureaucrats. But if the process is well designed and is sustained, it will become a useful tool to all departments and operations.

Who Should Plan?

Planning is too important to state administration to be left only to professional planners. Planning must involve people at levels of state government, starting with the governor. Experience from private corporations suggests that planning can only be successful with the support of the chief executive (Donnelly, 1980). If agency heads or the governor consistently ignore the planning process and the information made available by planners, then the process is useless. There will be little incentive to take the time to analyze options systematically if the results will not be used. George Michals, Executive Vice President of Genstar Corporation, argues that long-range planning is really not a grass roots exercise, but must start with those at the top of an organization clearly defining the goals and objectives of the organization (1979). Since the legislature has a significant role in the budget process, legislative leadership must also be committed to the planning process.

Some corporate planners have argued that line managers and chief executives are the only people who should do strategic planning since they are the only ones who have an overview of the corporation's strategy and they have the personal responsibility for carrying out the planning goals and objectives (Naylor, 1979). This view is too extreme. Junior staff often have more detailed knowledge concerning specific operations. The planning process should strive for a balance between centralized staff initiatives and grass-roots participation. But the vigorous and continued support of the governor in the planning process is vital to its success.

If planning is to assume a central role in influencing public capital investments, it is important to ensure that the people doing the job have experience of how agencies operate and understand the day-to-day problems faced by those managing public programs and facilities. As the manager of corporate planning for Dunlop has observed, "corporate planning . . . should be regarded as neither a profession or (sic) career in its own right . . . it should be an invaluable stage in the development of promising men and women who already have some relevant experience" (Rossiter, 1979, p. 20). The most successful corporate models are those in which many of those with line career ladders spend a period engaged in staff planning activities. "Planning," observes Edmund Burke on the White House Domestic Policy Staff, "is not viewed solely as a responsibility of a staff of experts. Indeed, the aim is to push planning down into the organization as much as possible. In this context, planning becomes a part of the process of administering an organization" (in Walter, ed., 1980. p. 1).

Agency heads should not expect their line staff to plan in their spare time. Individuals who have proved themselves able to run line operations well should be drafted for defined periods (of at least a year) to assist the agency head in planning. No more than two or three people are necessary at any one time and should report directly to the agency head. Even though planning will not be a permanent job for the designated planners, they should be provided with some training as they enter their term as staff planners. The training program may be developed by a local university and should include a debriefing session with outgoing planning staff. For large-scale projects or fundamental changes in the structure or mission of an agency, the planning staff may be supplemented by additional staff hired from within or even from outside the agency. It is not unusual for private corporations to subject some of their operations to a wide-ranging planning audit—often conducted with the assistance of outside consultants. The purpose is to inject fresh ideas and insights into the ongoing process.

The planning activities of state agencies must be coordinated. Trade-offs between projects occur not only among projects within given categories, such as rehabilitation of a road in one county rather than new road construction in another, but also among different categories, such as expanded prisons versus increased spending on health facilities. Interagency coordination of planning must be the responsibility of staff in the governor's office.

Several states have used blue-ribbon panels or commissions to establish broad capital planning goals and to establish the need for strategic planning (see Colorado, 1981). The panel is usually named by the governor and includes eminent members of the business community, labor, academia, economists, and representatives of state government. This process can be extremely useful in alerting the public (and government agencies) to the types of social and economic changes that can be anticipated, and to the need to increase spending on public works, and to under take strategic planning. The State of Colorado used a blueribbon commission to establish capital spending priorities. But the planning process cannot end with the preparation of a report. The capacity to sustain planning must be built into the operation of agencies and executive staff.

Setting Up a Capital Planning Process

The first step in establishing an effective planning process is to conduct a "planning audit" (Naylor and Neva, 1979). This involves the following four tasks:

1. *Review the present capital planning environment.* What agencies are engaged in pertinent planning and planning-related activities? What type of data concerning public works are being compiled and collected on a regular basis? How are the data analyzed? What links are already established among agencies? What links have been established between state and local governments with respect to public facilities planning?

2. *Specify the requirements from the planning process.* What types of reports and analyses are needed? What are the deadlines that these reports must meet in order to be useful for the budget cycle? Who is best equipped to prepare these reports?

3. *Define the goals and objectives of the capital planning process.* What functional areas are judged to be of greatest importance? What is the basic philosophy of the administration toward the allocation of responsibility between state and local agencies? Between public and private sectors? How should protecting the existing economic base (or existing capital facilities) be weighed against developing new enterprises and industries (or new capital facilities)? How should environmental protection objectives be weighed against economic development? To what extent is public facilities planning to be used as a way of controlling local growth patterns?

4. *Evaluate the effectiveness of existing planning resources.* Are the skills of those who are or will be engaged in planning appropriate to the needs? Are planning activities conducted in those agencies that can use them most effectively? Is there sufficient executive oversight over the planning activities?

This audit provides the information necessary to establish the ongoing planning process. For the process to begin successfully, the governor and staff must define, in some detail, the objectives of the planning process and how planning will proceed.

The format of the capital plan will depend upon the structure of the planning process and the resources available at the state level. It will also depend upon the distribution of fiscal and administrative responsibility for infrastructure between state and local governments. As a check-list, however, a plan should cover the following topics:

- *The overall objectives of state development policy.* The plan should specifically state the relative emphasis placed on economic growth, environmental protection, and the geographic distribution of development, for example.
- *A description of the planning process and participants.* A description is necessary so that those who must act on the plan and who may have information relevant to planners understand where the plan came from.
- *Long- and medium-term forecasts of major economic, fiscal, and demographic variables.* The forecast should include a section that outlines how these forecasts have changed since the previous plan was issued.
- *Detailed analyses, by category, of the levels of investment currently made needed in the future for public works.* These analyses should include annual operating, maintenance and repair costs, as well as investment in new projects. This section should include analyses of major alternatives and evaluations of the implications of reducing or increasing operating and maintenance spending.

- *Evaluation of alternative financing methods for major new projects.* This should include as assessment of the costs associated with each alternative financing option in order to determine the most efficient approach.
- *Assessment of the fiscal resources available for capital spending.* This is a long-term projection of the revenues produced by different taxes, fees, and charges.

Patience and persistence will be needed to develop an effective capital planning process that is suitable to the needs and resources of the state. Over time, the behavior of the participating agencies will change as a result of participating in this continuing process. Indeed, day-to-day decisionmaking and managing will also change. Hall (1979) examined how the management process in five major corporations altered as a result of introducing strategic planning. He found that they shifted from ad hoc decision making to decisionmaking that used the strategic planning process. This led to changes in the organizational structure—changes that could be adopted by state agencies as they incorporate planning in their day-to-day operations. These changes were as follows:

- Information collection and analysis tended to be organized more in terms of broad strategies and less in terms of specific projects.
- Centralized staff and top managers became more active in defining the organization's goals and strategies.
- Organizations were restructured around the broad strategies.
- A shift from rigid hierarchies toward the use of management by objectives.
- More frequent corporate reorganization to accommodate new strategies.

While the structure and level of effort devoted to strategic planning by state government will depend upon the resources and responsibilities of the state, these changes suggest the types of initiatives that will be necessary to formalize the planning process. Members of the governor's staff will have to work with staff designated by agency heads. They will have to delegate information gathering and analysis to agency staff familiar with data sources and analytic techniques. They will have to establish lines of communication with budget and agency line staff. They must be prepared to suffer through a painful learning process. And the chief executive must be prepared to use the planning process to establish priorities and evaluate alternatives and policies. But unless planning becomes an integral part of infrastructure policy, the crisis of poor maintenance, inadequate public works, and wasteful capital projects will remain.

Conclusions

Capital planning is not a mystical process that is conducted under the shroud of indecipherable economic models and reams of computer print-outs. The purpose of planning is to provide decisionmakers with well-organized and accurate information on the costs and benefits of alternative projects. Four characteristics of planning should guide the establishment of the planning process:

- planning is a continuous process,
- planning involves the management of information—both incoming and outgoing,
- planning is an amalgam of science and art, and
- initiating a planning process will meet with resistance.

Planning should not be carried out only by professional planners, but should be a logical step in the careers of those with line responsibilities. It is a staff function, and planners should report directly to agency heads and to governors. With the overt support of the chief executive, the planning process can become integrated within the decision-making and budget processes and will yield more than dust-gathering reports. A commitment to planning is not a commitment to a *plan* but to improving public investment decisions.

CHAPTER IV

The Economics of Planning and Managing Public Investments

THERE IS NO SHORTAGE OF PROPOSALS for repairing and developing the nation's infrastructure. Neglected maintenance in recent years makes almost every bridge, sewer, water supply system, and highway a candidate for upgrading. The growth and shift of population could justify the construction of numerous new facilities. Even in the best of times, limited budgets prevent public officials from pursuing all the infrastructure maintenance and investments that would benefit the public. In an era of shrinking federal funds and tight state and local budgets, deciding which projects to pursue and which to postpone or sacrifice is even more difficult. The stakes are large. These decisions will have a direct impact on the quality of life and the economic health of the entire community.

First, decisionmakers confront formidable obstacles in effectively investing in public projects. First, there are countless alternatives from which to choose. A dollar of state revenues could be spent repairing any cracked road, replacing any leaking waterpipe, or constructing any number of new facilities. Limited time and money bar the identification and consideration of all promising projects. If policymakers attempt to meet all the needs that have been indentified, they will face an impossible fiscal burden.

Second, the outcomes that are anticipated from a public investment decision may not be the eventual outcomes. No one defers maintenance of a bridge in the expectation that it will collapse. But some bridges inevitably fail. Many of the factors that ultimately determine the wisdom of a decision—such as energy prices, population growth, or the level of economic activity—are beyond the control and predictive ability of decisionmakers. Even when options are few and well defined, the uncertainty of the future makes it difficult for the policymaker to identify the best course of action.

Third, choosing among dissimilar outcomes, even when accurately forecast, is often difficult. How does one compare the benefits of a smoother ride on a resurfaced highway with a decline in infant mortality due to improved health facilities? Yet these are the type of trade-offs that must be made because state budgets are constrained—dollars spent on one project are at the expense of other projects.

Finally, decisionmakers in the public sector operate under handicaps not faced by their private sector counterparts. They often lack critical information that guides private sector decisions. For example, *market prices* indicate to businessmen the benefits and costs of alternative courses of action. For many goods produced or allocated in the public sector, however, market prices do not exist—clean air, cultural activities, and personal safety, for example. Even where public prices do exist, they may not reflect the true social benefits and costs. *Profits*, the excess of revenues over costs, provide businessmen with a readily measurable and generally reliable measure of how efficiently they are making decisions. Public investments are not intended to maximize government revenues, but to serve the public interest, for which there is no direct measure of performance. The absence of critical information places a greater burden on the judgment of public decisionmakers who must then rely more heavily on educated guesses and conjecture. Fortunately they can be aided by the application of established economic principles.

The subject matter of economics is resource allocation. Public policy decisions—state budgets and programs—are also concerned with resource allocation. The economic way of thinking provides a logical and practical approach to public policy decisions. When skillfully applied, the economic way of thinking simplifies and illuminates the alternatives of choice and produces more effective decisions.[1]

Anyone who has completed a course in economic theory may doubt that economics has anything to contribute to real-world decisionmaking. Conventional economic theory often fails to address the difficulties faced by decisionmakers. It assumes away the problems posed by inadequate information, complexity, and uncertainty.

This apparent contradiction evaporates, however, when one recognizes that economic thinking is not economic theory. An appreciation for economic thinking will lead public sector decisionmakers to the same conclusion reached by economist William Baumol (1976):

> I can say quite categorically that I have never encountered a business problem in which my investigation was helped by any specific theorem, nor may I add have I ever met a practical problem in which I failed to be helped by the methods of reasoning involved in the derivation of some economic theorem.

But if governors and their staffs are to use economic reasoning, they must understand the basic concepts. This chapter provides a brief discussion of the logic of economic choice and a brief guide to cost-benefit analysis.

The economic approach to decisionmaking is frequently identified with cost-benefit analysis, a formal technique for comparing the benefits of a public investment project (such as a highway, hospital or waste-disposal facility) with the cost of the project. Actually, cost-benefit analysis is only the formal application of the general economic procedure of weighing costs and benefits, which can be applied to any resource allocation decision, including maintaining, pricing, and replacing public facilities. The economic way of thinking does not reduce decisionmaking to the application of formulas. It never specifies what project ought to be financed or what policy should be pursued. It does help predict what outcomes will result from pursuing a particular course of action.

Economic thinking is a framework for decisionmaking—a way of thinking about the alternatives of choice. This framework produces insights into the nature of choice, exposes common errors in decisionmaking, and simplifies the comparison of complex alternatives. This chapter describes how economic thinking can be applied to planning and managing public works. The first section discusses the logic of economic choice, describing how the costs and benefits of a decision may be estimated. The second section examines how costs and benefits may be weighed directly to assess whether a specific project is a worthwhile investment of public funds. The final section summarizes the major conclusions.

The Logic of Economic Choice

Weighing benefits against costs is the economic criterion for decisionmaking. A potential course of action is attractive when the value of the benefits it will generate outweighs the costs incurred in taking the action. But to apply this economic criterion, one must recognize that economists' definitions of benefits and costs have very specific meanings which, like many economists' definitions, are often at variance with common usage.

The economic definitions of benefits and costs are well-suited for decisionmaking because they are the logical consequence of a simple assumption about the nature of choice. *Decisions are choices among competing opportunities.*

Benefits

Will an increase in transit fares generate additional revenue? Can replacement of deteriorating water mains be safely deferred? Will future population growth justify the construction of an additional high school? No matter how much relevant information decisionmakers collect, and no matter how carefully they plan, the benefits of many actions cannot be guaranteed at the time of decision. There is always some contingency—some unexpected event or some unanticipated outcome—that cannot be fully accounted for in advance.

A public works project yields many benefits that planners must attempt to measure. A new road will reduce the commuting costs of people who drive to work. It will also reduce the transportation costs of local firms. A wastewater treatment plant will improve the water quality in local rivers and lakes which will benefit recreational and commercial fishermen and people whose property is adjacent to these bodies of water. Consequently, the benefits for which a project is undertaken can be characterized as chances for gain or *opportunities.*

Costs

One consequence of choice is certain. Since the resources used in a project have alternative uses, the decision to commit them for a specific project means they cannot be employed elsewhere. Land used for a highway cannot be used for residential, industrial, or commercial development. Nor can funds spent on highway construction be committed to health care, mass transit, or education.

When resources are scarce, allocations are essentially trades: all the potential uses for a resource are exchanged or sacrificed for its chosen use. The pursuit of one opportunity always requires the sacrifice of other opportunities. This is one of many interpretations of the economist's oft-quoted dictum: "There's no such thing as a free lunch." The cost of a resource in any particular use is, therefore, the value of that resource in its best alternative use, or more broadly, *the cost of any decision is the value of the best alternative opportunity thereby forsaken.*

This definition of cost, known as the alternative cost or opportunity cost doctrine is probably the economist's major contribution to the practice of decisionmaking. No other economic principle provides as many insights or exposes as many errors.

Net Advantage

In a world of scarcity, choice is always a compound event. An opportunity is created by the commitment of resources to a specific purpose and others are simultaneously sacrificed by the withdrawal of resources from other potential uses. The potential advantage from choosing one course of action rather than another is therefore a *net* advantage—that is, the difference between benefits (opportunities created) and the costs (opportunities sacrificed).

Net advantage (or comparative advantage) is the economic criterion for decision-making. The most attractive opportunity is the one with the greatest net advantage.

When budget constraints prevent the state from investing in all the projects whose benefits exceed their costs, those promising the largest net advantage should be undertaken first. The comparison of competing projects is greatly simplified when the projects are evaluated—that is, when their consequences are expressed in numerical terms. This can be done by considering the marginal costs and marginal benefits associated with a specific project.

The logic of *marginal analysis* is simple. The marginal benefits and marginal costs of a decision are the incremental benefits and costs resulting from the decision. In choosing among projects, the proper question to ask is: What will change if a particular alternative is chosen? Only those things that change as a consequence of the decision need concern the decisionmaker.

Marginal analysis provides a simple rule for deciding whether to initiate, expand, contract, or terminate some activity: *do so only if the marginal benefits outweigh the marginal costs.* In plain English, it pays to do something only if what you expect to gain exceeds what you expect to give up.

Adhering to marginal analysis avoids making some common errors that result when these principles are ignored. Two major sources of error are: the belief that costs incurred in the past (sunk costs) should be weighed when considering present investment decisions; and, confusing average for marginal costs.

A common argument to justify a budget appropriation is that many millions of dollars have already been spent on a project and that failure to spend more money will jeopardize this prior investment. For example, a highway link may be urged because tens of millions of dollars have been spent constructing the highways to be linked. A $5 million extension to a convention center will be claimed to be essential if the state is not to lose the $75 million already invested in the center. These arguments are fallacious. Historical costs have already been incurred. Marginal costs are future costs—those that will be incurred only if a decision is taken and other opportunities are sacrificed.

Sacrifices already incurred, whether for productive or unproductive uses are *sunk costs.* They are irrelevant in evaluating a present decision. Only future changes in the levels of benefits and costs matter. If the marginal costs of a decision are expected to outweigh the benefits, the alternative should be rejected irrespective of what has already been spent. Throwing good money after bad may have emotional appeal but does not lead to prudent investments.

Average costs are easily computed but may yield misleading information concerning the viability of a project. Costs (or benefits) calculated as averages of past experience are accurate only if past and future circumstances and opportunities are identical. If unit costs vary with the scale of output (as the famous law of diminishing returns suggests), or if resource prices change, an average of past costs may be a very poor estimate of future incremental costs.

Another area where a reliance on average rather than marginal costs can lead to inefficient decisions is the pricing of public facilities or services. When the demand for a public facility is characterized by sharp fluctuations over time leading to high demands in peak hours, the use of average cost as a basis for price will lead to waste. Highways, mass transit, water, and electricity generation are common examples.

The marginal cost of providing service from a peak-load facility varies with the level of demand. The cost of adding 20 passengers to a half-empty train is virtually zero. The costs of adding another 20 passengers during rush hour are considerable, since they must include the capital and operating costs of an additional car.

If a single fare, based on the average cost of a ride, is charged at all times, the system will not be operating efficiently. During off-peak hours the fare will be too high, discouraging potential riders who could be provided the service at virtually no cost. At peak hours, on the other hand, the fare will be too low, leading to overcrowding. Off-peak riders will, in effect, be subsidizing peak period users. As a result, if it is economically feasible to vary fares according to demand, setting fares at marginal cost leads to the most efficient allocation of resources.

Project Evaluation

Roads, power plants, and wastewater facilities are costly and long-lived assets. Investing wisely in these assets has a high pay-off in terms of preserving the local quality of life, creating an environment conducive to economic expansion, and containing the costs of state and local government. The importance of these decisions warrants a formal economic analysis, specifying the nature of anticipated benefits and costs, and the basis on which their magnitudes are forecast.

Unfortunately, many public investments are made without the benefit of rigorous evaluation. A major public construction project is a highly visible sign of the local elected official's commitment to do something to revitalize downtown, to attract industry, or to create jobs. In the desire to establish this commitment, the utility of the finished structure may not be closely examined.

The typical sign next to the construction site promises a new public convention center that will provide 5,000 jobs and an additional $60 million in local taxes. It omits any reference to the annual cost of debt service, the operating deficit that will be met out of general revenues, how much it will cost to repair and maintain the structure, and how much the new convention activity will cost in added police expenditures, wear and tear on access roads, and congestion. There is, unfortunately, a mystical belief that public subsidies for economic development projects— whether hotels, convention centers or industrial parks— will be repaid many times over in jobs and taxes. Generous employment multipliers are used to provide a positive bottom line on projects that private developers have shunned for years. In fact, the jobs created by a project are a *cost* not a benefit, and the taxes paid as a result of the center should not be included as either a social cost or a social benefit.

The excessive attention lavished on job creation when evaluating public investment projects can lead to very poor decisions. British economic policy has included many major investments intended to create jobs—often at very high cost. Even if John DeLorean's risky automobile venture in Northern Ireland had been successful, each job created would have cost the British taxpayer $50,000. The goals of long-run job creation are attained most effectively by investing in those public works projects that meet rigorous cost-benefit criteria. If local transportation, water and other demands are met through cost-effective investments, the local economic climate will be strengthened.

While federal agencies were willing to provide subsidies to local projects in the name of urban renewal, neighborhood revitalization, or downtown development, all state or local officials had to gain from a more rigorous approach to project evaluation was the loss of federal funds. But as federal involvement declines, local projects will require local funds, and greater caution is inevitable. However, planners and well-intentioned officials must recognize that there is a strong tendency to undertake public investments for cosmetic reasons. After all, the benefits—construction jobs and highly visible ground breaking and ribbon cutting ceremonies—occur immediately, while the problems from a noneconomic project will only be experienced several years later. A rigorous evaluation process conducted by the state planning office will therefore be in conflict with the political interest of the executive office. A planning operation that consistently opposes projects favored by the chief executive will rapidly find itself excluded completely from the decisionmaking process. At the same time, an operation that is little more than a rubber stamp for politically inspired projects is contributing nothing to the development of an effective, public investment strategy and will inevitabily atrophy. There is no simple solution to this potential conflict.

The process of analyzing potential investments for inclusion in the capital budget is called project evaluation or benefit-cost analysis. "Benefit-cost analysis is ultimately a framework for policy decisionmaking, essentially a way to array the pros and cons of a programmatic decision together with a set of rules for weighing the importance of the pros and cons" (Gramlich, 1981).

The goal of project evaluation is to develop a ranking of investment alternatives in order that limited public funds can be spent as effectively as possible. Net present value (the net advantage of an investment opportunity) is the criterion for this ranking. The task of the capital planner is to identify the net present values associated with investment options.

Selecting from among these public investment opportunities requires more than economic reasoning. The personal judgment of the decisionmaker plays a large role. Alternative courses of action are identified and their consequences predicted, frequently on the basis of woefully inadequate evidence. Operational definitions of social benefits and costs must be developed, definitions that recognize the timing and riskiness of investments.

This section examines the practical aspects of making public capital investment decisions. It first discusses some of the problems in identifying and measuring social benefits and costs. Then it discusses the net present value decision criterion, and techniques for forecasting benefits and costs. This is not intended as a how-to guide to project evaluation, which can be gained from several comprehensive texts.[2] Rather, it presents the basic concepts to illustrate how capital planning can use economic techniques.

Social Benefits and Social Costs

No public capital investment benefits everyone. Every project creates some groups who gain at the expense of others. For example, the gainers from an urban freeway financed from general tax revenue include the users of the new road, businesses that construct the highway, and property owners who realize capital gains as a consequence of improved access. Losers include taxpayers who do not use the road, displaced residents who feel they were not fairly compensated for their property, and property owners who suffer a capital loss as a consequence of highway noise and air pollution.

How does one determine whether the gains to one group justify the losses to another? This is a political question, not an economic one. Economists avoid this problem by focusing entirely on efficiency and remaining silent on the question of equity. Economists claim no special insights into what is "fair" and therefore ignore the identity of individual gainers and losers. While the economic way of thinking identifies those who gain and lose, and the dollar value of these gains and losses, it does not declare whether these reallocations of wealth and income are justifiable.

A variety of benefits and costs can be expected from the construction of a new freeway. Benefits include reduced travel time and gasoline consumption as well as greater comfort and safety. Cost includes the sacrifice of the best alternative uses of the land, labor, and raw materials required to construct and maintain the road, as well as the clean air destroyed by the greater concentrations of automobile emissions.

The benefits of greater highway comforts and safety cannot be directly compared to the best alternative use of a road grader. To compare benefits and costs, a common denominator is required to reduce heterogeneous qualities to a single dimension.

The market value of the resources is the most obvious common denominator. For goods traded in competitive markets, price is a measure of the value of that resource in alternative uses. But market value has a number of defects as a common denominator of social values. The market price of a good is not a reliable indicator of its value to the user. For example, even where water is metered, no one pays as much for water as they would be willing to pay. Moreover, market values are determined not only by relative scarcity but also by the nature of property rights. If auto emissions damage health and property, the price of a gallon of gasoline will not measure the true cost of using it. External costs are omitted from market values.

Economists use money value as an index of value, but as a common denominator of value, they use willingness to pay or *use value* (utility). The benefits of reduced travel time or greater comfort are the dollar amounts that people would be willing to pay to achieve those ends. The cost of a construction worker is the wages another employer would be willing to offer to hire that worker. The social benefits and social costs of a project, therefore, are determined by answering the questions: How much would society collectively be willing to pay or sacrifice for it (benefits)?, and how much will it be required to pay or sacrifice for it (costs)?

Willingness to pay is a subjective state of mind that cannot be observed or measured. An objective measure of benefits and costs would certainly be convenient, but economists now recognize that no such measure exists.

Market prices are relevant to the valuation of benefits only insofar as they provide valuable indirect information about use values. Since a good will be purchased only if it is valued at more than its price, the price is an estimate of the minimum value placed on it by purchasers. Someone who pays $2 to ride a bus must value that trip at a minimum of $2. The value may be much more, but how much more cannot be directly inferred.

When market prices of benefits are unavailable it may be possible to derive a shadow or implicit price by measuring willingness to sacrifice for a similar activity. For example, the amount someone pays to park in a lot rather than park free and walk 10 minutes provides indirect information about the value the individual places on 10 minutes of his or her time.

In general, budgetary costs and prices are only the starting point for the estimation of social costs. Additions, deletions, and revaluations are generally required. Capital planners must apply considerable ingenuity in finding ways to use available data to estimate the value a community places on the benefits a proposed project may yield and the value of the resources that the project will use. Applied economics literature provides planners with many examples of applications of economic thinking to estimate benefits and costs. These are summarized in recent books about project evaluation (see, for example, Gramlich, 1981).

Net Present Value Criteria

The *net social benefits* of a capital investment are found by subtracting social costs from social benefits. But benefits and costs are often incurred at different times. How can two projects yielding different series of net benefits over time be compared? Consider two proposed investments, A and B, that have annual net benefits over four years forecast as follows:

Year	Net Benefits	
	A	B
Present	− 100	− 200
1	30	70
2	50	80
3	40	100

Both have total net benefits greater than the costs, but the certain costs are incurred in the present, and the uncertain benefits are not received for up to three years. Are either of these projects desirable? Which is more attractive? If funds were sufficient to finance one but not both of these projects, or if the projects represent alternative ways of meeting a local need, which should be chosen? The answer to these questions depends on the investment criteria employed. A variety of simple investment criteria are commonly used, such as selecting the investment that pays back most quickly, but these rules-of-thumb are often logically defective. Only one decision rule, Net Present Value, accounts fully and properly for the value of foresaken opportunities.

An investment is a commitment of resources in the present in the expectation of gains in the future. Yet the value, today, of benefits that will not be enjoyed until several years in the future is less than the "face value" of those benefits. After all, the fact that interest must be paid on savings accounts indicates that people must be compensated for delaying spending. Put another way, a dollar's worth of benefits received one year from now is not worth as much as a dollar of benefits available today. The later the availability of benefits, the smaller their present value. The present value of a stream of future benefits is today's value of all future benefits. The Net Present Value of an investment, therefore, is simply the present value of benefits less the present value of costs. It is a measure of the value created by the investment.

Discounting[3]

Since they are less valuable than current benefits, future benefits must be discounted to determine the total net present value of a project's benefits. The *discount rate* converts future dollars into an equivalent number of current dollars. The discount rate (sometimes called the capitalization rate) makes it possible to convert a stream of future values into a single value—the present value.

Discount factors for various rates of discount can be found in a variety of finance related books.[4] (They are also pre-programmed into financial calculators.) In using discount rates, keep in mind that a discount rate given to three decimal points may create a specious accuracy when multiplied by estimates of social benefits that have large margins of error.

Finding the appropriate discount rate, a critical element in any decision, will be discussed below.

An Example of A Net Present Value Calculation

The net present value decision rule is as follows: If proposed capital investments are not mutually exclusive, then any project with a net positive net present value should be pursued because the value of what is created is greater than what is sacrificed. If two projects are mutually exclusive, either because funds are limited or because it is physically impossible to undertake both projects, then the investment with the largest net present value should be pursued.

The net present values of the two hypothetical investment projects are computed for discount rates of 4 and 10 percent.

At a 4 percent discount rate, both projects are desirable, but B has a larger net present value (NPV) and would be preferable if only one project could be pursued. At 10 percent, both net present values are smaller. Project A has a negative net present value and is therefore not attractive. Project B is still desirable but less so at this higher interest rate. A higher discount rate indicates a smaller

Project A

Year	Benefit	4% Discount Rate		10% Discount Rate	
		Present Value Factor	Present Value In Dollars	Present Value Factor	Present Value In Dollars
Present	−100	1	−100	1	−100
1	30	.962	28.86	.909	27.27
2	50	.925	46.25	.826	41.30
3	40	.889	35.56	.751	30.04
Net Present Value			10.67		−1.39

Project B

Year	Benefit	4% Discount Rate		10% Discount Rate	
		Present Value Factor	Present Value In Dollars	Present Value Factor	Present Value In Dollars
Present	− 200	1	− 200	1	− 200
1	70	.962	67.34	.909	63.63
2	80	.925	74.00	.826	66.08
3	100	.889	88.90	.751	75.10
Net Present Value			30.24		4.81

relative value on uncertain future benefits and reduces the attractiveness of investments.

Selecting the Discount Rate

The magnitude of the interest rate used to discount future benefits is critical to efficient public investment. A small change in the discount rate can change the viability of a project dramatically. A survey made by the Water Resources Council of 245 authorized Corps of Engineers projects showed that, for about one-third of them, the costs would exceed the benefits if the discount rate were raised from the 5⅜ percent actually used to 7 percent (Anthony and Herzlinger, 1980). What is the appropriate rate for discounting future benefits into an equivalent amount of current benefits? Although economists have debated this topic for years, they have yet to reach a consensus as to how to measure this critical parameter.

The opportunity cost of public funds is what they could have earned in their best alternative use. The discount rate should measure this *social opportunity cost* of capital. Unless a project earns at least this rate of return, it will not cover the cost of the capital resources it uses. But this concept is difficult to measure. In choosing an interest rate, the effect of anticipated inflation must be taken into account.

The interest rate determined in the bond market reflects the anticipations of borrowers and lenders of future inflation. If inflation is 10 percent in the coming year, then dollars returned to lenders will be worth 10 percent less than the dollars initially received by the borrower. Under these circumstances lenders who wanted to increase their real purchasing power by 3 percent a year by lending, would have to receive an annual return of 13 percent.

The interest rate actually paid on bonds and other debt instruments is the *nominal* interest rate. The inflation adjusted interest rate—in this case, 3 percent (13 minus 10 percent)—is the *real* rate of interest.

If nominal interest rates are used in discounting costs and benefits of competing projects, then estimates of future benefits and costs should also reflect inflation. The simplest assumption is that the benefits and costs will increase at the same rate as the overall price level or, perhaps, at the same rate as a relevant component of the overall price index, such as construction costs. If the discount rate used is the real rate of interest, then benefits and costs should be estimated in *current* dollars and not be allowed to grow with inflation. The measurement of benefits and costs and the discount rate should reflect consistent assumptions about inflation.

Government Discount Rates. Since benefits and costs will extend over a long period of time, the choice of a discount rate is especially important for public works projects and will influence which projects are selected. The smaller the discount rate the greater the relative attractiveness of longer-lived and capital intensive projects.

In 1972, the Office of Management and Budget issued Circular A-94 which established a uniform real discount rate of 10 percent (with a few exceptions such as the post office and water-resource projects). Most economists believe this rate is much too high and that a *real* discount rate of 3-4 pecent is a realistic measure of the opportunity cost of funds.

Risk

For bonds promising identical payments, risky bonds sell at a lower price than do investment grade bonds, and investment grade bonds are cheaper than U.S. government bonds. The pattern of prices in financial markets provides compelling evidence to support the belief that, collectively, people are risk averse. That is, they prefer certain gains to riskier ones. In financial markets, investors trade reduced risk for a lower rate of return and willingly accept a smaller rate of return on insured savings for the peace of mind that their capital is secure.

Preference for safety over risk has direct implications for project evaluation. Every opportunity has two relevant dimensions—the expected net gains and the riskiness of those gains. Not only the expected net benefits of a project but the riskiness of those benefits should be taken into account.

All projects promising a 10 percent rate of return are not alike. A project that is almost certain to yield 10 percent is more valuable than one with a 50 percent chance of yielding 20 percent and a 50 percent chance of yielding nothing. (In technical jargon, it is not only the expected return but also the *dispersion* of the expected return that matters.)

Risk assessment creates serious problems for decisionmakers. In practice, there are no reliable techniques for taking risk into account. The extensive theoretical literature on risk is largely irrelevant because even the simplest measure of risk—the standard deviation of expected returns—cannot be forecast with any degree of accuracy.

There are two common procedures for grappling with risk in project analysis: risk adjusted discount rates and sensitivity analysis. Neither is wholly satisfactory, but they have the virtue of making explicit the often neglected consequences of risk.

Risk-Adjusted Discount Rates. Since people are risk averse, they will only undertake riskier investments if they offer a higher return than a secure investment. In reality, there is not one, but many social rates of return. Discount rates can be adjusted for risk as well as for the passage of time. For example, if a risk discount of 6 percent were added to a time discount of 4 percent, the present value of project A in the example above would fall from 10.67 to −1.39. When risk is taken into account, a proposed capital investment that was previously attractive may no longer appear so.

The problem with using risk-adjusted discount rates is determining the appropriate discount rate for a particular project. For projects without historical precedent, forecasting risk is largely a matter of guesswork. Predicting the distribution of educated guesses may be pure speculation. Estimating the appropriate discount rate is bound to be subjective and the conclusions disputable. Nevertheless, for risky projects, such as those using untested technologies or serving nascent markets, the error of ignoring risk is likely to be larger than that of imperfectly accounting for it.

Sensitivity Analysis. An alternative method for accounting for risk is to examine the underlying factors that determine the levels of benefits and costs. The range of uncertainty may be narrowed by computing the present value of a project using the "best case" and "worst case" scenarios. A project that has a positive net present value under a reasonable worst case scenario is going to be a stronger contender than one that switches from a positive net present value under favorable assumptions to a negative net present value under less favorable conditions.

Forecasting

Forecasting the future level of benefits and costs is among the most critical judgments required in project evaluation. A small change in some critical variable, such as the degree of utilization of a facility, can produce dramatic changes in the projected net present value of a project. The failure of projects to live up to expectations can usually be traced to errors in the forecasts of a critical component of benefits and costs. At best, the past provides indirect, incomplete, and uncertain predictions about the future. Projects may be so novel that information on which to base forecasts is non-existent. For other projects, information may be available but too expensive to collect.

Since planners are not usually specialists in forecasting techniques, they frequently rely on expert forecasts provided by government agencies or firms specializing in the relevant types of forecasts. It is worth noting that an assessment of the performance of expert forecasts raises questions about whether they are more reliable than those decision-makers could provide using less sophisticated forecasting techniques (see Ascher, 1978). Another problem with reliance on expert forecasts is that experts commonly disagree, leaving the decisionmaker with the difficult task of deciding which expert to trust.

Forecasting Techniques

Forecasting techniques are of two types: subjective and objective. Subjective methods are also called "implicit, informal, clinical, experienced based, intuitive, guesstimates, WAGs (wild-assed guesses), or gut feelings . . . the critical thing is that the inputs are translated into forecasts in the researcher's head. . . . Objective methods on the other hand, use formal processes that can be replicated by other researchers" (Armstrong, 1978, p. 387). Some objective methods for providing the long range forecasts required by planners are as follows:

- *Extrapolation.* When the process being forecast is not well understood, it may be most reasonable to simply assume that the trend that has persisted during recent years will continue in the future.
- *Curve Fitting.* Some phenomena, such as population growth or diffusion of innovation, can seemingly be described and predicted with a curve. The critical question is what type of curve is appropriate.
- *Econometric Models.* Variables to be forecast may be included in a mathematical model of the entire economy or sector thereof. Coefficients of the model are estimated using historical data and these coefficients are used to forecast future magnitudes.

In many contexts, objective forecasting methods are preferred because they minimize the role of personal opinion. There is, however, a substantial body of evidence to suggest that when changes in variables are small and few objective data are available, subjective methods are superior (Armstrong, 1978).

Limits of Forecasting

No reputable demographer forecast the baby boom following WWII or the baby bust of the 1960s, nor did economists predict the rise in oil prices and the inflation of the 1970s. The future is never fully predictable and all forecasts are subject to error. In assessing the riskiness of public works projects, it is essential to recognize this margin of error.

In general, forecasting errors will be larger the more limited the forecaster's knowledge of the forces at work. The limited knowledge of weather patterns, for example, makes it difficult to forecast floods and droughts. The accuracy of the forecasts will also decline the further into the future one predicts. There is a momentum of events in the immediate future that does not extend into the distant future. Short-term forecasts will, in general, be more reliable than long term. One must also conclude that a carefully considered and debated judgment may prove much more useful than the detailed output of a complex econometric model.

Conclusions

The role of the capital planner is to prepare analyses that allow policymakers to understand the relative merits of alternative public investment projects. Rational decisionmaking requires standards for evaluating various options, and a set of directions for applying these standards. The economic approach outlined in this chapter provides a reasonable and logical framework for comparing complex policy choices. These objective standards, however, must be applied within the legal, political, and institutional constraints discussed in the previous chapter. The economic approach allows reasonable projections of resource movements and revaluations that determine who would gain and who would lose and the expected size of these gains and losses that result from a public project. It also helps identify the policies and projects that achieve their objectives with the smallest possible sacrifices.

There are limits, however, to what can be achieved through economic analysis. Although economists can venture educated guesses about the dollar value of gains and losses, they can never conclude that the gains to one group, no matter how large, justify the losses to another group, no matter how small. This would require interpersonal comparisons not of dollars, but of satisfaction—outcomes that cannot be measured. Consequently, economic analysis alone should never be used to justify the selection of one project over another. Every public investment affects both efficiency and equity: the size of the pie and who gets the slices. Economists consider only questions of pie size, not who gets the pieces. Economic reasoning may determine whether a measure is efficient but it can never determine whether it is fair.

Governments and markets are two alternative mechanisms for allocating resources. Which is more efficient depends on the technical nature of the particular good or service (see Chapter 5, below). Efficient production occurs when the appropriate quantity is produced with the minimum possible cost. This will happen when the benefits and costs measured by decisionmakers represent the true social values. In this case, self interest and the marketplace can be relied upon to induce decisions in the public interest. In some notable circumstances, however, the incentives facing decisionmakers are such that the operation of the market will pass up projects for which social benefits exceed social costs.

There are some goods, such as clean air and public safety, that are consumed *collectively*, and for which it is difficult to charge beneficiaries an efficient market price. Assessing the value of these goods is extremely difficult but not impossible. Without government intervention, these goods will not be efficiently produced. Air pollution results from the failure to charge polluters for the valuable disposal services of the air mantle. As a result, the social cost of polluting activities such as driving—which includes damage to the environment—exceeds the private cost faced by drivers. When private markets lead to inefficient decisions, the government can improve resource allocation by either providing goods directly or by altering marketplace incentives through subsidies, taxes, and regulations. These programs face many problems, however, because measuring or predicting benefits and costs and designing practical programs are difficult.

The application of the principles outlined in this chapter can help improve the quality of the information that is made available for public infrastructure investment decisions. Yet, improving state decisionmaking will not necessarily lead to efficient investments. Overlapping or ill-defined responsibilities among federal, state, and local programs lead to poor decisionmaking. Strings attached to federal aid distort local priorities. Failure to apply appropriate financing mechanisms leads to poor management and maintenance. And the politics of public works often leads to pork-barrel projects. These problems only reinforce the importance of developing more rigorous planning and project selection procedures.

CHAPTER IV NOTES

1. For an introduction to the economic way of thinking see Paul Heyne, *The Economic Way of Thinking*, (Science Research Associates, Chicago, 1983), and Israel M. Kirzner, "The Economic Point of View," (Sheed and Ward, Kansas, 1960).

2. For a comprehensive treatment of the methodology of project evaluation see Gramlich, 1979.

3. The principle of discounting is quite simple—logically, it is simply the inverse of compounding. For example, $100 compounded at 5 percent will be worth $105 one year from now and $110.25 two years from now. On the other hand, $110.25 received two years from now or $105 received one year from now will be worth only $100 today if the discount rate is 5 percent. To compound at 5 percent, the present value is multiplied by a compounding factor of $(1.05)^t$, where t is the number of years of compounding. To discount at 5 percent the future value is multiplied by a discounting factor of $1/(1.05)^t$ where t is the number of years in the future the payment will be received.

4. See Brealy and Myers, 1981.

CHAPTER V

Assigning Responsibilities: Public or Private?

POLICYMAKERS MUST DETERMINE how responsibility for financing, constructing, and operating facilities should be allocated between the public and private sectors. The lines of demarcation have changed over time. On the one hand, the experience with large-scale energy projects in western states has led some private developers to assume the responsibility for building facilities that are traditionally the responsibility of state and local governments—such as schools, water systems and roads (described in Financing Public Works, Chapter 6). On the other hand, some states have financed the development of industrial plants and facilities—traditionally a private responsibility—for the "public purpose" of creating jobs. During the past decade, the share of the revenues from the sale of tax-exempt state and local bonds that is devoted to traditional public infrastructure has declined from 75 percent to less than 50 percent (see Vol. 2: Financing, Chapter 5).

Declining revenues and tight budgets are causing a growing number of states and localities to reassess what activities they can afford to finance. For many fiscally pressed jurisdictions, privatization has become an important policy issue. Yet it is also a very complex issue. Privatization may mean abandoning a public service or facility and turning over the responsibility for service delivery to private firms, or it may only mean maintaining public sector control over the development and operation of a facility, but transferring ownership to a private corporation to take advantage of federal tax incentives.[1] The purpose of this chapter is to explore the economics of these different forms of privatization. The first section describes the complex interaction between the public and private sectors in the provision of infrastructure, and how traditional distinctions have become blurred over time. The second section outlines the economic reasons for public provision and control of facilities. The third section discusses the different ways in which private sector resources can be used to assist in the provision of infrastructure. These include complete privatization, cost sharing, contracting out, and leasing. The fourth section provides some general guidelines to privatization. The final section presents the major conclusions from the chapter.

Public and Private Responsibilities

The borders that separate public and private sector responsibilities in the construction and maintenance of infrastructure are separated by a wide no-man's land—contested terrain where the financial burden has shifted with changing perceptions of the roles and objectives of public policy. Some state and local governments have been willing to use public monies to pay for the construction of industrial plants, wastewater treatment facilities, transportation connections, and even equipment to be used by private firms in order to attract or retain jobs. Federal grants from the Economic Development Administration and the Department of Housing and Urban Deveopment have been used to cement these "public-private partnerships" where local constitutional and statutory constraints would otherwise have prohibited the use of local public resources. Hotels have been subsidized, convention centers built by state and city agencies, and sports stadia erected with public dollars. North Dakota has a state-owned bank. The City of Green Bay owns the local football team. And New York City operates 17 hospitals.

On the other hand, the problems associated with boom-towns resulting from large-scale energy projects, and curbs on local taxing and spending powers have led other states and localities to require private developers to finance traditionally public infrastructure such as schools, roads, and water supply systems. Some observers believe that turning over responsibility for public services as diverse as mass transit and education would lead to more efficient and lower cost service delivery.[2]

These conflicting developments have not proceeded uniformly in all states. Tradition and legal constraints have restrained the advance of the public sector in some states, while stalled growth and eroding fiscal bases have propelled it forward in others. Cuts in federal grants and the delegation of responsibility from the federal to state governments are likely to slow the growth in the use of public funds for traditionally private purposes and are encouraging some jurisdictions to develop innovative ways to finance and operate public services and facilities.

Privatization is a complex issue. It is not a simple either-or choice between public or private ownership and management of a park, a waste disposal facility, or a bus fleet. There are many degrees of private and public control. A distinction can be drawn between control of production and control of consumption. For example, there are many commodities whose production and ownership are in private hands, but that are subsidized in order to encourage a level of output above one that would result from a "pure market" outcome. For example, residential housing is, for the most part, privately produced and privately owned, yet homeownership is subsidized through the personal income tax because it is thought to bestow benefits to society beyond those enjoyed by the homeowners themselves. Rental housing for those with low incomes is subsidized through rent assistance payments. In other cases, private producers of a good or service are subsidized. Shipbuilding is conducted privately but with huge direct and indirect subsidies because protecting the industry from foreign competition is argued to be in the interests of national security. The development of synthetic fuels is aided through low interest loans, loan guarantees, or guaranteed markets.

Other industries, such as electricity generation and distribution, are conducted under private ownership but are regulated as natural monopolies. In education and health, services are provided by heavily regulated, and usually nonprofit, institutions that compete directly with public institutions. In these cases, public involvement often reflected the desire to extend the services to those who could not afford to pay for privately supplied services. Railroads were privately developed but with public land grants. Almost all inland water transportation systems are publicly provided, while the users are private, for-profit companies. In some cities, garbage is collected by private firms under contract to the city, and in other cities it is collected by municipal employees. For any given type of public facility—from a bridge to a water supply system—the choice is not between public ownership and management and private ownership and management. It is a choice over the appropriate type and level of public intervention.

Market Failure and Public Intervention

Markets and governments are alternative mechanisms for achieving the gains that social cooperation makes possible. Market coordination is characterized by voluntary exchange and decentralized decisionmaking. Government decisionmaking, in contrast, is centralized, and coordination is achieved through a hierarchical chain of command. The degree to which governments allocate resources varies widely among sectors and states.

In the production of most goods, the spontaneous operation of the marketplace is more effective than the deliberate coordination of the government. The prices generated by voluntary exchange provides producers and consumers with the information and incentives that governments cannot produce. There are, however, important exceptions to this rule.

While economists are generally impressed with the free market as a coordinating mechanism, they recognize its limitations. There are circumstances where government intervention in the marketplace can be expected to improve the allocation of resources. These circumstances are known as market failures and include public goods, externalities, and natural monopoly.

- *Public (or Collective) Goods.* Due to their peculiar nature, some highly desirable goods cannot be produced by voluntary cooperation. To enjoy a meal in a restaurant people must pay for it. To benefit from police protection, people need not pay for it. A meal is a "private" good, and those who do not pay for it are excluded from consuming it. Police protection, in contrast, is a "public" good that is, the benefits of police protection are shared equally and simultaneously by all people whether or not they pay for it. Individuals will not offer to pay for a good they expect to get for nothing if someone else pays for it. They will attempt to be "free riders." If everyone attempts to ride for free, no one rides at all. Consequently, public goods, such as police protection, clean air, public parks and mosquito abatement, must be provided (i.e. financed) by government or they will not be produced in sufficient quantities. The feasibility of excluding those who will not pay is the key property that determines whether a good can be produced in the private sector.

The "public" in public goods refers to its consumption properties. Not all publicly provided goods are public goods. Most education is publicly provided, but education is not a public good—those who do not pay for it can be excluded. If that were not true, private schools could not exist.

● *Externalities.* An externality (also called a third-party, or spill-over, effect) occurs when voluntary exchange affects parties not directly involved in the exchange. A firm that dumps toxic waste in a neighborhood landfill imposes external costs on residents. Homeowners who upgrade the external appearance of their home, thereby increasing property values, bestow external *benefits* on their neighbor.

Since decisionmakers do not take external effects into account, too much of a good that generates external costs and too little of a good that produces external benefits will be produced. Basic education provides external benefits. The nation as a whole benefits from having a literate population.

When externalities exist, resource allocation may be improved by government intervention in the marketplace. By taxing or restricting activities that produce external costs and by subsidizing activities that produce external benefits, social welfare can be improved. This result can occur only if the government has the necessary information and can provide effective incentives, and if the cost of administering the program is smaller than the anticipated gains.

- *Natural Monopoly.* The technology of producing some goods is such that they are most effectively produced or distributed by a single supplier. It would be wasteful, for example, to have firms compete in the supply of water. Since the protection of the consumer from paying excessive prices provided by competion will not be forthcoming, the government takes direct responsibility by providing these goods directly or by regulating the firms that do.

A second justification for public intervention is concern over "equity." A service or facility may be publicly provided and owned to ensure equal access to all consumers regardless of income. Public education—at the primary and secondary school levels—is a clear example. Redistribution can be achieved either through direct public provision of the service or through subsidies to low-income individuals (perhaps in the form of vouchers). Concern with equity, however, may extend beyond merely ensuring that everyone has sufficient financial resources to pay for services. It may extend to enforcing an egalitarian provision of services, constraining the ability of those with enough money to purchase better quality services. Opponents of tax credits for tuition costs in private schools have made this argument.

Market failures and equity considerations cannot explain all the activities currently undertaken in the public domain. It seems doubtful the public would approve of the "privatization" of prisons, even if a private firm offered to incarcerate inmates at a lower cost and to use them more productively than was possible at public facilities. Pride of community ownership may be a "psychic good" for which voters are prepared to pay through taxes. Baltimore's Aquarium, Green Bay's Packers, or New York's Public Library may be valued less by the local community if they were privately owned, even if there were no changes in their outward appearance and operation. These considerations impose real, if elusive, constraints on the extent to which public responsibilities can be turned over to the private sector.

But a strong case for public intervention does not necessarily require that the full responsibility for financing and delivering the commodity or service rests with the state or local government. It is obvious that the same results in terms of service costs and quality can be achieved through an almost infinite combination of public and private ownership of facilities, responsibility for service delivery, regulation, and indirect tax incentive or subsidy. The costs of successfully administering different systems, however, may differ markedly, and the efficiency of alternative means of intervention may also vary. The following section examines the experience with alternative methods of "privatization" and attempts to establish some guidelines for more effective assignment of responsibilities between public and private sectors.

Shifting the Balance

The variety of arrangements used by local governments for service delivery defy easy generalizations. A recent survey conductd by the International City Management Association (ICMA) examined the different mechanisms used by city and county governments to ensure the delivery of a wide range of public services—Table 2 summarizes their results. They also draw some general conclusions concerning the effectiveness of different methods that provide some general guidelines for state and local governments. Their applicability to a particular government will, clearly, depend upon legal and constitutional constraints, the political and economic environment, the strength of public employee unions, and many other factors.

Contracting Out

Undertaking a formal agreement with a private (for-profit or nonprofit) firm is widely used for commercial solid waste disposal (41 percent of governments), residential waste disposal (34 percent), and tree trimming (30 percent). The local government pays the private firm to

Table 2
Service Delivery Approaches of Cities and Counties

SERVICE	NO. OF CITIES AND COUNTIES REPORTING	LOCAL GOVERN-MENT EMPLOYEES		INTERGOV-ERNMENTAL AGREEMENTS (%)
		IN PART (%)	EXCLU-SIVELY (%)	
Public works/transportation				
Residential solid waste collection	1,390	12	48	8
Commercial solid waste collection	1,143	24	28	7
Solid waste disposal	1,314	14	35	29
Street Repair	1,640	33	65	5
Street parking lot cleaning	1,483	11	84	3
Snow plowing/sanding	1,282	19	79	4
Traffic signal installation/maintenance	1,569	37	53	14
Meter maintenance/collection	767	7	72	3
Tree trimming/planting	1,454	39	53	4
Cemetery administration/maintenance	718	16	68	4
Inspection code enforcement	1,588	14	82	6
Parking garage operation	784	16	73	7
Bus system operation/maintenance ...	555	20	24	38
Paratransit systems operation/ maintenance	579	28	19	26
Airport operation	561	25	37	24
Public utilities				
Utility meter reading	1,204	23	64	8
Utility billing	1,248	25	62	9
Street light operation	1,281	21	30	20
Public safety				
Crime prevention/patrol	1,659	22	74	5
Police fire communication	1,685	16	75	14
Fire prevention/suppression	1,520	18	69	8
Emergency medical service	1,361	27	39	16
Ambulance service	1,256	19	30	16
Traffic control, parking enforcement ...	1,502	7	90	4
Vehicle towing and storage	1,310	14	7	2
Health and human service				
Sanitary inspection	991	13	49	33
Insect/rodent control	1,059	22	44	27
Animal control	1,508	16	61	18
Animal shelter operation	1,262	13	36	28
Day care facility operation	441	19	7	15
Child welfare programs	567	37	26	26
Programs for elderly	1,190	57	18	21
Operation/management of public elderly housing	611	21	20	41

	CONTRACTING							
PROFIT (%)	NEIGH-BORHOOD (%)	NON-PROFIT (%)	FRAN-CHISES (%)	SUB-SIDIES (%)	VOUCHERS (%)	VOLUN-TEERS (%)	SELF-HELP (%)	INCEN-TIVES[1] (%)
34	0	0	15	1	0	0	1	0
41	0	0	17	1	0	0	1	0
26	0	2	5	0	0	0	0	0
26	0	1	0	0	0	0	0	0
9	0	0	0	0	0	0	0	0
14	0	0	0	0	0	0	0	0
25	0	2	1	0	0	0	0	0
5	0	0	0	0	0	0	0	0
30	1	1	1	0	0	3	3	0
10	1	8	1	1	0	3	1	0
6	0	1	0	0	0	0	0	0
11	0	2	2	1	0	0	0	0
21	1	8	4	8	0	1	0	1
22	2	20	4	13	2	7	4	0
21	0	4	9	2	0	1	0	0
9	0	1	10	0	0	0	0	0
12	0	1	8	0	0	0	0	0
38	0	2	14	0	0	0	0	0
3	5	2	0	0	0	9	5	0
1	0	3	0	0	0	2	0	0
1	1	3	0	1	1	17	1	0
13	1	10	3	5	0	15	0	0
23	1	9	4	7	0	14	0	1
1	0	1	0	0	0	1	0	0
78	0	0	7	0	0	0	0	0
1	0	5	0	1	0	0	0	0
13	0	5	0	0	0	0	1	0
6	0	8	1	1	0	0	0	0
13	1	17	1	3	0	2	0	0
33	6	34	2	15	2	4	3	2
5	2	22	1	8	1	6	2	0
4	4	28	1	13	3	18	7	1
12	1	17	0	4	0	1	0	2

Table 2 (Cont.)

SERVICE	NO. OF CITIES AND COUNTIES REPORTING	LOCAL GOVERN- MENT EMPLOYEES IN PART (%)	EXCLU- SIVELY (%)	INTERGOV- ERNMENTAL AGREEMENTS (%)
Health and human services (cont.)				
Operation/management of hospitals ...	393	8	16	21
Public health programs	743	35	25	28
Drug/alcoholic treatment programs	635	30	10	28
Operation of mental health/retardation programs facilities	508	25	13	32
Parks and recreation				
Recreation services	1,458	39	51	9
Operation/maintenance of recreation facilities	1,539	35	58	8
Parks landscaping/maintenance	1,574	20	76	5
Operation of convention centers, auditoriums	452	15	68	9
Cultural and arts programs				
Operation of cultural arts programs ...	707	46	11	11
Operation of libraries	1,189	20	48	26
Operation of museums	505	25	21	15
Support functions				
Building grounds maintenance	1,669	25	73	4
Building security	1,499	11	85	3
Fleet management vehicle maintenance				
Heavy equipment	1,642	37	59	2
Emergency vehicles	1,560	34	59	3
All other vehicles	1,622	32	63	2
Data processing	1,471	23	64	10
Legal services	1,605	29	41	6
Payroll	1,719	11	86	2
Tax bill processing	1,320	23	64	10
Tax assessing	1,098	14	54	27
Delinquent tax collection	1,254	16	59	18
Secretarial service	1,656	5	94	1
Personnel services	1,663	8	90	2
Labor relations	1,514	25	69	3
Public relations information	1,547	12	87	1

Note for Tables 2 through 13: Percentages when total may exceed 100% because respondents indicated more than one service delivery approach.
[1] Regulatory and tax incentives.

	CONTRACTING							
PROFIT (%)	NEIGH-BORHOOD (%)	NON-PROFIT (%)	FRAN-CHISES (%)	SUB-SIDIES (%)	VOUCHERS (%)	VOLUN-TEERS (%)	SELF-HELP (%)	INCEN-TIVES[1] (%)
25	1	24	1	4	1	2	0	1
7	2	25	1	8	1	7	2	0
6	4	38	1	12	1	6	2	1
6	3	38	1	15	1	5	1	1
4	5	12	2	4	1	19	5	0
8	3	9	0	1	0	4	1	0
9	1	2	0	1	0	4	1	0
5	1	6	3	1	0	2	1	0
7	7	38	2	17	2	31	6	0
1	1	10	0	6	0	11	1	0
3	3	30	1	16	0	20	2	0
19	0	1	0	0	1	0	0	0
7	0	1	0	0	0	0	0	0
31	0	0	0	0	0	0	0	0
30	0	0	0	0	0	0	0	0
28	0	0	0	0	0	0	0	0
22	0	2	0	0	0	0	0	0
48	0	2	0	0	0	0	0	0
10	0	1	0	0	0	0	0	0
22	0	2	0	0	0	0	0	0
6	0	4	0	0	0	0	0	0
10	0	3	0	0	0	0	0	0
4	0	0	0	0	0	0	0	0
5	0	1	0	0	0	0	0	0
23	0	1	0	0	0	0	0	0
7	0	2	0	0	0	1	0	0

Source: Martha A. Shulman, "Alternative Approaches for Delivering Public Services," *Urban Data Service Reports*, Vol. 14, No. 10, Washington D.C., International City Management Association, October 1982.

provide the service. The procedure is most efficient for services and facilities with the following characteristics:

- Services being offered for the first time, so that no public employee lay-offs are involved. Lay-offs can be avoided, however. To avoid lay-offs when it contracted for private waste disposal, the City of Phoenix, for example, requires successful bidders to offer positions to all displaced city workers, although they are not required to keep them if their work is not acceptable (*Public Management*, October 1982).

- Services whose output and cost can be readily calculated and compared with public sector performance (e.g., tree trimming, pot-hole filling, etc.). Some cities maintain some public employees engaged in waste disposal in order to monitor the relative cost of private contractors.

- Services for which many competitive bidders exist so that the government does not fall victim to monopoly pricing (e.g., legal services, data processing, and computer programming).

- Services for which the demand may fluctuate unpredictably.

- Services for which local public demand is insufficient to merit the purchase of expensive equipment, for example, the city of Phoenix contracts out for chip-sealing the streets because the equipment is very expensive. A private contractor can use the equipment on a year-round basis by serving several jurisdictions.

The National Conference of State Legislatures recently surveyed state and local officials on the issue of contracting out public service (Kirkland, 1982). While most respondents agreed that much more could be done, they also identified the barriers to increased contracting out. The most serious barriers were those over which the state has no control—inertia and public employee resistance.

If the state wishes to facilitate contracting out, one of the most useful steps it can take is to provide local governments with technical assistance on how to undertake the process—providing model contracts, training on cost monitoring and accounting techniques, and pooling information from areas that use contracting.

Ted Koldene (*Public Management*, October 1982) summarizes the issue:

> The readiness of the community for alternative approaches will be an issue, one way or another. In some communities, the city employees may have an unbreakable lock on service production; in others, some commercial vendor may have a lock on the work. The media may by sympathetic, or may view a proposed change as some kind of rip-off. And, perhaps most important, there is the question of alternative producers. Nothing happens unless there are other willing and able parties to assume service delivery (e.g. volunteer, or mutual-help organizations, contractors available for bid). When these alternative service producers are not present, they will need to be cultivated. Each community is different. Some will have a fairly substantial ability to implement alternatives, others will not. The ability to redesign and to adopt the service delivery system is a kind of resource—its almost like having a favorable natural climate.

Franchise Agreements

A franchise grants a private firm the exclusive or non-exclusive right to provide a service to a given geographic area. By issuing franchises to private firms for parking or waste handling, for example, the city or state can reduce the need to make public investments in facilities or equipment. Franchising differs from contracting in that the consumer or user pays the private firm directly and the local government plays a regulatory role, usually setting prices and specifying service levels. It can be applied to services with the same type of characteristics as those that lend themselves to contracting and for which the users can be clearly identified and the services can be readily evaluated by those consumers. Waste collection, ambulance services, and parking garages are examples of services for which franchises are frequently used. There is a fine line between granting enough franchises to ensure competition and granting so many that franchisers cannot enjoy economies of scale, therefore operating inefficiently.

Subsidies

Subsidies are financial or in-kind contributions from government to a private service supplier or to consumers of that service. The approach is used most frequently for transit, mental health services, and cultural services. Subsidies can therefore reduce the need for capital investments. Subsidies can be allocated as a lump-sum payment or can be allocated as material, equipment, labor, or rent-free space in a local government building—either for general purposes (to improve library services) or for specific purposes (to increase the number of books in a library). Subsidies are appropriate when a direct peformance contract is not feasible because the local government cannot anticipate the level of services that will be needed (transit) or because the services must be tailored to the needs of individual clients (mental health). The cost savings arise because the public subsidies "leverage" private expenditures but must be carefully monitored because the area may not be competitive and because nonprofit firms, the most frequent recipients of subsidies, do not respond to profit incentives.

Vouchers

A voucher is a subsidy to a service user, usually provided to those with low incomes or specific disabilities, given in the form of a nontransferable certificate. It gives the consumer freedom of choice while fulfilling the financial responsibility of the government. For the voucher to encourage efficiency in production, there must be both competition in the service-supplying industry and ease of entry into the industry. The administrative costs may be high, particularly if the vouchers are given out on the basis of income or other measures of need. These costs should be carefully monitored.

Regulatory or Tax Incentives

Regulatory incentives can be used in a variety of ways that reduce the need for government activity—from requirements that garbage be placed on the curb to relaxations in licensing and accreditation rules for day care centers to encourage private suppliers. Tax incentives are, in effect, similar to subsidies and can be used as widely, for example, to encourage home care for the disabled or elderly or to encourage private firms to provide day care facilities. Both regulatory and tax incentives are complex to administer and difficult to monitor.

The ICMA survey found a marked difference in patterns of privatization among cities of different sizes. Large cities (populations above 250,000) were much more likely to contract out for most public services (except for day care and hospitals) than small cities (populations below 10,000). The failure of smaller cities to take advantage of the private sector in this way is probably due to the difficulty of setting up and monitoring contracts and to the lack of competition among potential suppliers. State governments may be able to play an important role in providing technical assistance to localities in setting up procedures for contracting and evaluating contractors' performances.

Privatization of service delivery is only one way in which the need for public financing of facilities can be reduced. The second is the transfer of ownership of facilities to private owners to take advantage of both the tax benefits (unless Congress revokes them) and the improved administrative and management efficiency that experienced developers and profit-seeking firms can offer. Unfortunately, although there are many claims of the lower costs associated with the private construction and management of facilities, there are no reliable data to substantiate the level of savings that can be realized. In the absence of any evidence, the following section identifies some of the issues that must be addressed in transferring ownership from state or local governments to private corporations.

Tax Subsidy Financing[4]

In general, tax subsidy finance involves several different yet related ways that state and local governments may capture federal tax subsidies designed ostensibly for private capital investment. With the passage of the Economic Recovery Tax Act of 1981, Congress and the President put into law subsidies for private capital investment that are deeper and more explicit than those in previous law. Even with some of the reversals found in the 1982 Tax Equity and Fiscal Responsibility Act, the subsidy to business capital investments is still strong.

If the ownership of the facility is in private hands, the private owner may get a subsidy more valuable than the taxes due. As a result, the transfer of ownership from public to private hands, at least for tax purposes, may reduce costs. There are various ways in which ownership, for tax purposes, may be transferred:

- *Safe Harbor Lease.* New York's Metropolitan Transit Authority (MTA) sold the tax advantages of the ownership of subway cars and buses to a private firm. The MTA got a cash payment related to the present value of tax advantages of ownership. The firm, treating the payment as a cash investment, realized a fair after-tax rate of return from the tax consequences of ownership.
- *Service Contract.* Several towns in Massachusetts, in conjunction with a private firm, have arranged for the private financing of a resource recovery facility. The facility will be owned and operated by the private firm; the towns sign a contract guaranteeing a minimum annual fee for handling municipal refuse. The fee paid by the towns under contract may be lower than the direct expenditures that would be necessary if the towns were to construct and maintain a comparable public facility.
- *Tax Exempt Leveraged Lease.* In a widely discussed deal, the City of Oakland sold a public museum and auditorium to private investors and leased them back for continued public operation. The cash received was sufficient to renovate one of the facilities and the City hopes to establish a fund sufficient to repurchase the facilities at the end of the lease term.
- *Outright Sale.* The State of New York has proposed that the World Trade Center in Manhattan, owned and operated by the Port Authority of New York and New Jersey, be sold to the private sector. Because of the tax advantages of owning real estate, the positive cash flow from the complex may be worth more to a private owner than to a public owner. The private sector may therefore be willing to come up with a purchase price that is worth more in present value terms to the public sector than the

tax free cash flows from the complex if it were to remain public.

Each of these arrangements can generate a public sector fiscal dividend, but each has its own legal, fiscal, theoretical and political problems (see Vol. 2, Financing, Chapter 7).

Some Considerations

Eight considerations should be weighed by any government before entering into any attempt to privatize a public service or facility. Cost should only be one factor among many in making this important decision. Privatization is not necessarily attractive simply because it is cheaper.

1. *How important is public control per se?* On a scale ranging from little to total public control of a facility, contracting out and outright sale lie at one extreme (high private control), and safe harbor leases (private ownership on paper only) or vouchers are at the other end. Leasing and franchising fall in the middle, depending on the terms of the lease or franchise. A jurisdiction may find a fiscal bonus in selling off a very public facility like a bridge, but must be prepared to treat the bridge operator as a regulated utility if it wants 24-hour access or certain maintenance standards. Public control of the facility or service is, of course, far less important with respect to, say, the World Trade Center, which is basically a real estate operation, or a resource recovery facility for which regulatory issues may not be critical.

2. *What kind of risk does the public sector want to run?* Some of the transactions are quite complex. The private franchise may have to be carefully scrutinized. For some lease agreements, areas of the tax law are relatively untested. What happens if the lease calculations are wrong and reversion of public ownership is not possible? What happens if the private franchiser's performance cannot be effectively monitored? Can government be held hostage in the future as a result of decisions made to avert a fiscal pinch today? Many lease transactions contain provisions to indemnify the investors in the event of legal

73

changes or adverse IRS actions. Such provisions create the risk, however small, that government may at some point be handed a huge bill by the investors in the amount of their lost tax advantages. Similarly, the state government may be liable for the actions of a private contractor.

3. *How does the conversion of a public facility or service to a private one affect the existing structure of state and local finance?* New York State, for example, requires that the principal component of any debt service payment not exceed by more than 50 percent the principal component of any other debt service payment from the same obligation. Unlike a home mortgage, state law requires local debt to be paid off early, and level debt service payments are not possible.

Refinancing a sewer through an Industrial Development Bond (IDB), however, allows for level debt service on the new debt, violating the intent of the state law. Similarly, private financing, like lease purchase arrangements, may provide a way around constitutional limitations on debt issuance.

4. *How will the state's standing in the bond market be affected?* Changing institutional behavior affects a state's credit rating. For example, in turning over facilities or equipment to private hands, it is turning over capital assets and revenue resources. If its contracting is not conducted efficiently, the cost of service delivery will not necessarily be reduced. In addition, creative financing may create new debt instruments that add to the jurisdiction's fiscal load.

5. *Who benefits from the fiscal dividend received?* Privatization does not necessarily allow a state to make greater investments in public infrastructure. If a government refinances an existing facility or privatizes a service, the effect, more often than not, is one of loosening up future budgets. There is no guarantee that savings will be used for capital investments. Savings can be spent to hire more police, teachers, or to cover a newly discovered general fund deficit.

On the other hand, there are ways savings can be seen as more or less supplementing the public capital investment process. If a specialized unit of government whose sole emphasis is infrastructure, say the Port Authority of New York and New Jersey, refinances a facility, its fiscal resources would be increased, and while it could hire more planners instead of more police and teachers, its likely emphasis will be on capital spending. The benefits of privatization for infrastructure investment therefore depend upon the agency administering the privatized facility or service.

6. *Fiscal Prudence.* Another concern is whether the flexibility afforded by privatization may allow for imprudence. The plain vanilla world of traditional government finance has standards to assure the long-term soundness of financial systems. Innovation creates hazards as well as flexibility. For example, a municipality, in establishing a fund to reduce local tax assessments, might be tempted to draw on the fund more heavily early during a lease term, thus passing on hidden increases to later years. No standards or laws exist to limit this type of borrowing against the future. Obviously, if all financial managers and officials were irreproachable, there would be no need for standards in this area. But since they are not, transactions must be carefully scrutinized and standards developed.

7. *In-State Financial Incidence.* Creative financing arrangements impose some hidden costs. Not all of these are borne by out-of-state taxpayers. Depending on the relationship of state and local tax systems to the federal tax system, some portion of the cost may be shifted back to within-state taxpayers. Savings enjoyed by a city may be paid for by state taxpayers. If the state and local government were judged to forego revenues through the issuance of tax-exempt debt, the loss at both levels would be considerably higher. The incidence of costs as well as benefits implicit in many privatization arrangements should be analyzed with care.

8. *Construction Costs.* Because of detailed guidelines that regulate the way in which state and local governments enter into and monitor contracts, a private firm may be able to build a facility quicker and cheaper. Unfortunately, there are no data that allow a detailed comparison of the potential cost savings, although there are claims that they may exceed 25 percent.[5]

Any proposal to turn over the construction or operation of a public facility or the delivery of a public service to a private firm should be analyzed with great care. Private is not necessarily cheaper or better than public.

Conclusions

After a century of almost continuous expansion, the last few years have witnessed significant retrenchment of federal, state, and local government activity and employment. Community centers and libraries are being closed. Residential developers are being required to build their own streets and sidewalks. Transit fares are being raised. Program cuts are being made in the same ad hoc fashion that prevailed during expansion. In making these retrenchment decisions, the political strengths of constituencies has been a more important consideration than whether a service is better conducted by the private or public sector. The Reagan Administration has accelerated the process by cutting back or terminating many categorical grant programs, and by converting others into block grants. "New federalism" could precipitate an even more radical surgery of programs and activities by state and local governments.

These changed circumstances call for a new strategy; states can no longer deal with periodic budget crises by across-the-board cuts in all programs. Further cuts in many programs will seriously jeopardize their usefulness. Nor can they expect, as in the past, to maintain programs by vigorous lobbying in Washington. Instead, state and local governments must determine which services can be turned over to the private sector or terminated. Only by redefining responsibilities will states have sufficient funds for those activities that need direct state government involvement.

Assigning priorities for public spending will require a rigorous examination of current practices. If a convention center is a demonstrable boon to a local economy, why is it not financed privately? Are public subsidies to private business necessary? Must a recreation site be developed and operated with public funds? Is public ownership of the water supply system necessary when electricity and gas are privately provided? Does an airport need to be owned by the county?

The basic guideline for state and local officials to follow in deciding who does what is that the public sector should only intervene if market-determined outcomes are either inefficient or inequitable or both. Even if public intervention is required, it does not have to involve spending public funds or public ownership of facilities. Regulation of private suppliers of drinking water, for example, can lead to the desired supply of services without full public ownership. The task is by no means easy, but without cutting back on some public services, there will not be enough tax revenues to pay for those services that properly belong in the domain of state and local government.

CHAPTER V NOTES

1. The tax advantages of leasing are threatened by tax proposals in Congress that would remove the benefits bestowed by the 1981 Economic Recovery Tax Act.

2. See, for example, the proposal to privatize New York City's subway system, and other proposals described in Savas, 1982.

3. This section draws heavily upon the excellent report prepared by Martha Shulman, "Alternative Approaches for Delivering Public Services" published by the International City Management Association (*Urban Data Service Reports*, Vol. 14, No. 10, October 1982).

4. This section draws upon a paper published by the Council of State Planning Agencies, written by Jeffrey Apfel, ". . .", and prepared for the HUD Governmental Capacity Sharing Program.

5. See remarks prepared for the Legislature of Georgia by Harvey Goldman, Arthur Young and Company, 24 September 1982.

CHAPTER VI

Capital Budgeting

CAPITAL BUDGETING IS THE PROCESS by which long-term plans are tied to annual or biennial state expenditure decisions. The capital budget is an essential element to efficient investments in and management of public works. Multiyear capital plans are not, in themselves, budgets. They are guides to identifying current and future fiscal year requirements. They provide decisionmakers with a perspective longer in range than expense budgets and therefore can show the consequences of today's budget decisions on tomorrow's economic and fiscal environment. The capital budget is an authorization to undertake long-term projects. Expense budgets are annual appropriations of the funds necessary to conduct state activities.

Capital budgeting essentially is a subset of strategic capital planning activities. It is a "control device used to carry out the various actions and operating plans arising out of the strategic plan" (Connelly, 1980, p. 6).

Few state and local governments prepare adequate capital budgets and even fewer effectively integrate capital plans with annual budgets. A November 1982 survey by the U.S. General Accounting Office (GAO) concluded that few states

- assess the effect of a public works project on future operating budgets;
- compare the net present value of all future costs with future benefits of alternative projects;
- properly account for costs and benefits attributable to a project;
- evaluate alternative methods of financing the project; or

- include all projects within an overall capital investment plan, and evaluate the implications of not making the investment.

Many states that have separate capital budgets fail to perform these basic functions. A separate capital budget should not be regarded as a useful end in itself. The separation of capital from operating expenditures is only an accounting exercise. But a capital budget that is an integral part of the planning and budgetary process and that provides decisionmakers with more timely and relevant information than is provided by current budgetary practices is an essential element to an effective state capital investment strategy.

A capital budget allows state and local governments to evaluate the condition of their public infrastructure, and how that condition is affected by public investment decisions. A private corporation that attempted to balance its books by ignoring depreciation on its plant and equipment would attract the attention of the Securities and Exchange Commission (SEC) and would invoke the ire of its stockholders. Yet that is precisely what all state and local governments are doing each year. Most have balanced their books and even financed tax cuts by underinvesting in infrastructure because they do not include a basic element in the accounts of all private firms—depreciation. A properly prepared capital budget would provide this missing element.

The federal government has failed to set an example for other levels of government. The GAO labelled federal capital budgeting as a "collection of haphazard practices." An audit of the problems of the U.S. General Services Administration (GSA), which oversees federal capital projects, illustrates the problems that impede efficient capital budgeting at the state level. The GAO concluded the following:

- Capital planning requires both long- and short-range plans, but the GSA only plans one budget cycle ahead. Few states prepare long-run plans that are used as part of the capital budgeting process.
- Capital investment decisions must be made in conjunction with overall departmental and program needs, but

the GSA makes decisions on a project-by-project basis. Again, most states do not prepare comprehensive capital plans that span all agencies and projects.

● Capital construction activities require links between planning and budgeting; GSA has failed to establish these links. In most states, plans are not closely tied with budgets, nor are the processes that lead to the preparation of plans and budgets integrated.

● Effective facility management requires a high priority to be placed on operation and maintenance activities; GSA's huge maintenance backlog reflects the lack of attention that has been given to these activities. Similarly, states are not always as attentive to maintenance and repair needs as they are to new construction needs.

These failings can be overcome by some relatively straightforward changes in planning and budgeting processes. Chapter 4 described how projects can be rigorously evaluated through the application of cost-benefit techniques. Planning issues were discussed in an earlier chapter. This chapter examines three aspects of establishing a capital budget. The first section discusses how to link capital expenditures to changes in annual operating budgets. The second section describes techniques to assess the condition of local public works in order to respond, more effectively, to the need to maintain and repair facilities. The third section analyzes how to integrate long-term plans with annual budgets. The final section summarizes the major conclusions.

Capital Spending and Operating Budgets

The first step in developing a rational and useful capital budget is to distinguish capital from operating expenditures. Capital expenditures are nonrecurring outlays associated with social, economic, and human capital investments—spending that will enhance the productive capacity of the state and its residents into the future. There are well established, generally accepted accounting principles for making these distinctions.

The second step is to identify the effect of today's capital expenditures on tomorrow's operating expenditures. If lawmakers realize how much a new convention center, vocational education institution, or expanded port facility will cost in terms of increased annual spending on maintenance and repair (operating expenditures) and in terms of the increased level of activities of other state agencies, they might pause before approving it. This is an essential element in estimating the net present value of the project (see Chapter 3).

The budget should also contain estimates of the impact of not undertaking major rehabilitation and repair projects on future operating budgets. This will reveal the consequences of deferred maintenance and encourage viewing maintenance and repair over a longer time horizon.

A major barrier to the adoption of rational planning and budgeting procedures at the state level is the way in which federal capital-aid programs are designed. Without a change in federal programs, state and local actions cannot overcome the strong biases built into many federal programs. The destructive influence of federal programs on state and local capital planning and budgeting was recently documented by the U.S. General Accounting Office. Year to year changes in federal program budgets, allocation procedures, and regulations deter long-term planning at the state and local levels. Indeed, the availability of federal support for certain categories of capital investment shapes state and local planning. The GAO documents how the highway program has caused shifts in capital spending that may not reflect local priorities:

> Many state officials believe that they must get and spend all the federal highway dollars for which they are eligible. Thus, they tend to plan highway programs around the federal-aid categories from which they can receive funds. As a result, what often happens is that a state, which may need to replace bridges, may instead build another section of interstate road because it has already spent all its money in the bridge replacement category and funds may still be available to build a new section of interstate (U.S. GAO, November 1982, p. 25).

A study by the American Society of Planning Officials (1977) cites examples of how federal funds influence local choice. It describes one city that awards points in its capital improvement rating system for the use of outside funding sources, and another city that plans projects so that full utilization is made of intergovernmental revenues. These practices are the rule rather than the exception. "We were also told of the possibility of states intentionally allowing their roads to deteriorate until they reach a point where federal funds would be available for major rehabilitation" (GAO, November 1982, p. 27). A study by the Department of Commerce concludes, even more bluntly, that "the Federal-aid highway programs have set the stage for deterioration of the nation's highways by subsidizing state and local construction and not maintenance" (1980, p. III.25). Wastewater construction grants exhibit the same bias.

Harry Hatry, of the Urban Institute states:

> Local governments are frequently faced with the option of constructing a new interceptor sewer that requires a maximum of 25 percent of local funds for funding sewer line replacement projects that typically require 100 percent local funding (1981, p. 50).

When Congress defines the categories for which federal funds may be spent, they are contributing to poor planning and the inefficient allocation of resources for public capital investment. Significant improvements in state and local capital planning and management are unlikely without changes in the way federal funds are allocated. In 1982, several bills to establish a federal capital budget were submitted by Congressmen from both parties, and subject to some qualifications (including the treatment of defense expenditures), the GAO recommended that the federal government distinguish between capital and operating expenditures for the annual budget. This would be a modest first step toward a more rational approach. A longer run solution will require the further decategorization of many federal programs. The past two years have seen the conversion of some federal categorical grants into block grants. The process will have to be extended and the experience of these recent initiatives carefully monitored. There are still major federal grant programs that do limit the type of investments that are eligible for federal support, including the interstate highway program, the rapidly waning wastewater treatment program, and mass transit support. The broadening of project eligibility may require a reduction in the share of projects that can be financed through federal aid.

Separating capital and operating spending and estimating the implication of capital investment projects for future operating expenditures are two important steps in using the capital budget to translate the capital plan into fiscal reality. The third step is to evaluate alternative ways of financing projects. The State of Califorinia systematically considers alternative financing techniques as it evaluates projects. For example, for a general purpose office building, the State determines the life-cycle costs of lease, lease purchase, and purchase options (see *Financing*, Chapter 7). Different financing methods can significantly affect the cost of capital, the use of the facility, and the efficiency with which it is managed (see preceding chapter).

Assessing Capital Maintenance Needs[1]

Any capital facility requires upkeep. Yet, few state and local governments systematically prepare estimates of the costs and consequences of failing to undertake needed maintenance. The result is that decisionmakers do not have any real information delineating the consequences of deferring vitally needed rehabilitation and repairs. Much of the present public concern over infrastructure has arisen because of the awesome dollar figures of needed investments that has come as a result of adding up estimates of needs and because of the uncertainty attached to these figures.[2] When standards are applied to determine needs, they are engineering standards that have not been subjected to the type of cost-benefit analysis outlined in Chapter 4.

The New York City Planning Commission in a 1979 report, concluded that:

> The frequent caution of 'garbage-in means garbage-out' is particularly applicable to capital planning. The City had to rely on . . . generally accepted engineering principles as

the basis for determining the useful economic life of its capital improvements. Though these provide general guidance, the Commission believes that the City must develop capacity to evaluate the needs of each individual element of its capital plant (1979, p. 49).

In a recent review of state and local practices conducted by the Urban Institute, Harry Hatry reached the following conclusions (1981, p. 3):

- Procedures exist for *assessing the condition of the infrastructure* in many service areas such as roads, bridges, sewers and wastewater treatment. These generally have not been widely tested nor adopted by local governments for regular condition monitoring. The assessments that are done are often undertaken as part of federal requirements.
- Procedures for *replacement analysis* (i.e., comparisons of replacement vs. rehabilitation vs. maintenance options, including examination of the full cost implications, the effects on level-of-service, the uncertainty and risks, and financing options for individual infrastructure components) are increasingly being recommended by outside professionals and by many federal, state, and local personnel. Such procedures have been given occasional trials, but there is little regular provision for them in local governments, except where required as part of federal grant regulations.
- Formal processes for *ranking and prioritizing capital choices among project proposals and service areas* are becoming more frequently used in local government. These rankings, however, tend to be highly subjective, appearing more objective than they actually are. If and when better information from condition assessments and replacement analysis is forthcoming, this information could greatly enhance the meaningfulness of the ranking/prioritizing process.
- *The technical procedures* for condition assessment and, particularly, replacement analysis are still largely untested by local government. Conceptually, these procedures make a great deal of sense. Their accuracy, their costs, and applicability to local government infrastructure analysis, however, have not yet been determined

adequately. Full-scale testing of these techniques in a local setting is needed.

- *Federal mandates* can have major impacts on local government choices, affecting costs (both capital and operating) as well as timeliness and service quality. Systematic considerations of federal and other funding sources is seldom done but is highly desirable.

- *Organizationally,* the capital infrastructure choice process is highly decentralized, with specific choices and project ranking performed primarily at the agency level. Decisions on capital projects are often made separately from the operating budget process (although this seems to be slowly changing). Seldom, if ever, do administrative control offices require or provide ground rules for systematic, comparable, condition assessments, or replacement analysis by operating agencies.

To assess capital maintenance and operating needs, planners and agency heads need to manage the type of data they compile and analyze. The International City Managment Association (ICMA) has developed a list of 36 indicators for decisionmakers (Groves, 1980). The 36 indicators are developed from 12 factors that affect the financial and economic viability of a city. These factors are classified as either environmental, organizational, or financial. For example, the level of intergovernmental revenues would be a financial factor, management practice would be organizational, and the condition of local capital facilities would be an environmental factor. The indicators of the condition of the capital plant are (1) the maintenance effort, (2) the level of capital outlays, and (3) the percent depreciation in the value of capital assets. By examining trends in these indicators over time, signs of significant deterioration and trouble spots can be identified. Capital budgeting, however, cannot rely on these indicators as a final measure of needed maintenance and repair expenditures. The indicators can be used to identify facilities and systems that should be examined more closely. The final determination of replacement, repair, and maintenance programs should be based on a combination of factors, including structural and mechanical data, and social and political considerations integrated through a cost-benefit analysis. It may cost $10 million to restore a road to a given standard, but it may not be worth $10 million in terms of reduced travel time, reduced vehicle maintenance costs, and improved safety to rehabilitate the road. Greater efficiency in establishing efficient capital maintenance programs will require some changes in the strings attached to federal aid programs—some of which establish rigid standards rather than performance or output measures.

A much less formal, but nevertheless informative, approach was taken by the City of Fellunde, a small city in Colorado: a detailed survey of future infrastructure demands was published in the local paper. Citizens were invited to respond and assign priorities for public works projects and were provided with detailed information on the costs of alternative projects. This does not do away with the need for in-house planning but it can encourage greater public awareness of the issue and provide planners with some indication of local demands.

The cost for needed investment in research and development to establish more efficient and effective maintenance and repair procedures is considerable. The cost can be minimized and the results disseminated more rapidly if the federal government participates with states in the process. States, in turn, will have to work closely with city and county governments to ensure that they are able to develop or adopt consistent and effective procedures. Techniques to monitor the condition of the local public capital stock and to estimate the costs and benefits of alternative repair and replacement procedures are urgently needed. Recent work by the Urban Institute illustrates that this can be done relatively inexpensively and can yield substantial cost savings and improvements in conditions over the life of the facility (see also O'Day and Neumann, 1983).

Translating The Capital Plan into a Capital Budget

The translation of a long-term capital plan into annual budgets is not an easy process. There will be an inevitable tension between those responsible for preparing the capital plan and the annual budget, just as there is a natural conflict between corporate planners and financial managers in the private sector (Schmidt, 1979). Budget officials are concerned with immediate fiscal constraints and details of projects. Planners compare costs and benefits in the long term.

The art of capital budgeting, and the responsibility of the governor, is to bring together planners and budgeters in a way that harnesses the abilities and perspectives of both. It is the art of successful conflict resolution, not of conflict avoidance. The data available to budget officers on the performance of programs and revenue sources should be integrated into planning activities so that plans include full consideration of costs and benefits. But conflict can only be resolved successfully if the chief executive takes a strong interest in and control of the planning process (Naylor, 1979). The emphasis, again, is upon establishing a process not upon selecting a given plan.

Capital budgeting is not easy nor is it free. The joint consideration of capital and operating budgets and regular condition assessment and replacement analysis are highly desirable but difficult procedures to establish. Many technical issues, such as establishing how to measure condition and formalize consistent cost-benefit analyses, have been better developed and tested for local government use.

The GAO describes the procedures through which the City of Baltimore—which they judged a successful organization—ties its capital planning and budgeting operations:

> The City of Baltimore requires its departments to determine, for each of its proposed capital improvement projects, the effect on the city's operating budget. The City's planning department, when reviewing department submissions, looks carefully for capital projects that will result in reduced operating and maintenance costs. Proposed projects may be cut if the operating budget is insufficient to finance the projects' operations and maintenance activities. A statement of how proposed projects will affect the operating budget is included for each project in the City's capital budget (November 1982, p. 53-5).

By contrast, the City of Cleveland's capital budget is entirely separated from its operating budget and as such does not consider the operating costs of any capital projects. The city's capital improvement plan has not been used in developing its capital budget. Six procedures can contribute to successful coordination of long-term capital planning and budgeting:

- The plan should include estimates of long-term costs and benefits of alternative capital investments.
- The plan should use capital budgeting techniques (such as life-cycle costing) where they are applicable and practical, so that decisionmakers and those preparing the annual budgets can judge the project's influence on annual operating expenditures.
- Select projects should be based on need rather than on the availability of funds.
- Projects should be selected for inclusion in the annual budget in a logical sequence from their multiyear plans.
- Funding mechanisms should be designated to protect the funds allocated for priority capital projects.
- The status of ongoing physical capital projects should be reviewed continuously to ensure that previously established targets (time, money, scope) are being met (U.S. GAO, 1982, p. 54ff).

Overall, the GAO study found that the governments that met their public capital investment and maintenance needs most successfully were characterized by extensive planning:

> Successful organizations prepare master (long-range) plans that link these master plans and mid-range (multiyear) plans with annual capital and operating budgets; and consider the long-term effects of planning on operations, maintenance, and capital assets (1982, p. 26).

Conclusions

There is nothing magical about integrating well designed capital plans into successful capital budgets. A capital budget allows policymakers to determine the fiscal implications of individual projects and to be able to see what steps are being taken toward reaching the objectives defined in the plan and, as important, to see the consequences of delaying a project or of deferring maintenance. The capital budget involves four basic steps:

1. Separation of capital and operating expenditures for the budget year;
2. Analysis of the fiscal impacts of proposed projects on operating expenditures in future years;
3. Evaluation of alternative methods of financing; and
4. Evaluation of the impacts of past projects.

In addition, repair and replacement procedures for capital facilities and programs must be developed that reflect economic costs and benefits—not merely engineering replacement standards. Finally, integrating the capital plan with the capital budget requires the design of a process whereby the inevitable conflicts between planners and budget officers can be resolved effectively. Background analysis and information are an important part of this conflict resolution, but so is the leadership of the chief executive. Capital budgeting will not prove useful without the direct intervention of the governor.

CHAPTER VI NOTES

1. The Urban Institute is undertaking an extensive HUD-sponsored project on techniques for assessing the condition of infrastructure directed by George Peterson and Harry Hatry. Marshall Kaplan at the University of Colorado is surveying many states on the condition of their public works. The National League of Cities (April 1983) has prepared an initial assessment of needs. At the state level, the most extensive effort has been undertaken in Washington by the Department of Community Development.

2. Dr. Pat Choate and Susan Walter report that nearly $3 trillion will be needed during the next decade simply to arrest the rate of decay. The Congressional Budget Office (1983) places the total much lower—an additional annual expenditure of $55 billion.

Selecting the Management Mechanism

PUBLIC FACILITIES CAN BE MANAGED under several different administrative structures, including state or local government agency, public authority, and special district. Each mechanism has advantages and disadvantages that makes it more suitable than others for working with certain types of public facilites. For example, public agencies are highly responsive to annual changes in executive or legislative priorities, but are less able to pursue longrun investment strategies. Public authorities are able to protect their longrun investment programs but are less responsive to executive and legislative oversight and may not coordinate their activities with the programs of other agencies. This chapter outlines how these alternative institutional or management mechanisms can be used to improve the planning, budgeting, and managing of public facilities.

The first section of this chapter discusses the importance of selecting the appropriate financing mechanism to provide guidance in planning, budgeting, and managing and to encourage efficient use of public facilities. The subsequent section outlines the concept of a state infrastructure bank and examines how this mechanism might be used to make more efficient use of federal, state, and local fiscal resources for facility development and maintenance. The final section analyzes the advantages and disadvantages of state agencies, public authorities, and special districts as mechanisms for managing public facilities.

Managing public facilities is the practice of carrying out decisions as effectively as possible and is concerned with the day-to-day running of public facilities. Both planning and managing are concerned with decisionmaking and with managing flows of information. Three characteristics of planning and managing that distinguish the two activities as they pertain to public capital facilities are as follows:

1. *Time Horizon.* Planners tend to be concerned with decisions that do not have to be made immediately or whose effects will not be felt immediately, for example, analyzing the demand for additional water storage facilities, or for an additional prison. The manager of a facility is more concerned with immediate decisions, such as hiring additional staff or repairing a roof.

2. *Degree of Detail.* Planners are not involved with such detailed decisions as managers. They may analyze the size and design of a new state water treatment facility, but they will not decide who is hired, or from where materials are purchased.

3. *Extent of Trade-Offs.* Capital planning involves comparing the values of capital investment projects among competing state agencies and authorities and even among different levels of government. If a capital budget were no more than a "wish list" of all the projects desired by different agencies, it would be far beyond what even the most well-endowed state could afford. Managers are much more circumscribed in the trade-offs they can make because they often work within fixed budgets that allocate funds to major expenditure categories.

Since some of the information generated by the operation of public facilities bears directly on the planning of future facilities, planning and management processes should interact. Good management involves regular evaluations of the operation of a facility and analyses of potential innovations in processes and procedures. These reviews will generate practical ideas about the capacity, the design, and the maintenance of future projects. The process should work both ways. Planners, as they evaluate alternative ways to meet future demands, may uncover information of immediate use to managers.

Choosing the Financing Mechanism

Selecting the financing mechanism for a category of public works is critical to efficient budgeting and managing. An adequate and responsive revenue source can cut the cost of financing and provide information that assists both planners and managers. For example, changes in the flows of revenues raised by a user fee can provide a vital indication of whether the level of service or the capacity of the facility should be changed. Charging users for access to a service or facility will also encourage more efficient behavior by users. This section defines the principles that should guide the selection of the financing mechanism.

State and local governments can turn to three primary sources of funds to finance the construction and maintenance of public capital projects. First, they can borrow, and service the debt either from general revenues or from dedicated tax revenues or user fees. Second, they can finance the projects directly out of tax revenues or charges and incur no debt. And third, they can require or encourage the private sector to incur at least part of the costs. There is no single efficient or equitable way of funding a public works project—what may work in one county may be constitutionally prohibited or institutionally impossible in another. This does not preclude establishing some general principles to guide an effective and equitable financing strategy. These principles apply regardless of the institutional structure selected. Because of statutory and constitutional restrictions, however, certain types of financing mechanisms may be precluded for certain types of administrative structure.

First, the pursuit of economic efficiency dictates that, wherever possible, those who benefit from a public facility should pay for its development and operation, and the amount they pay should be related to the level of their use. This is not always possible through a simple user fee. It may require the imposition of special taxes on beneficiaries (a gasoline tax to finance highways and roads, or a property tax surcharge for retail establishments near a subway stop, for example). Earmarking specific revenues is precluded in some states and is also resented by some legislatures and executives because it reduces annual flexibility. It is the excessive exercise of budget flexibility, however, that has created the chronic problem of underinvestment in public works.

Second, the cost of a public capital project should be amortized over the life of the project. This principle naturally follows from the first, since it ensures that there is no intergenerational transfer of net benefits or net fiscal burden. If a long-lived project is financed out of current revenues, then future generations will enjoy the benefits with none of the costs. Conversely, financing a facility with an expected economic life of 7 years with 20-year bonds will shift the cost to future generations and leave the jurisdiction with a future tax burden higher relative to this fiscal and economic base than at present.

Third, future operating and maintenance expenses associated with a project—operating a convention center, maintaining a bridge, or repairing and upgrading a resource recovery center, etc.—should be explicitly considered when planning and budgeting a project. If earmarked revenue sources are not sufficient to cover both debt service and ongoing expenses, unanticipated and often large subsidies from general revenues will be necessary.

Fourth, fiscal and administrative responsibility for a public investment project should be limited to those jurisdictions where most of the impacts are experienced. To illustrate, a program for maintaining roads and highways should be financed by a statewide tax, say on gasoline or vehicle registration. If county governments were individually responsible for their roads, then the transportation network would be a fragmented patchwork of roads that did not advance the state's economic interest. On the other hand, a waste disposal facility that serves only one to two counties should be financed and run by those counties. In this instance, the state may serve a regulatory function to ensure compliance with uniform environmental standards. Adhering to this principle reduces the tendency to finance pork-barrel projects at the state level, for example, voting for a state-financed convention center in one city in return for a state-financed industrial park in another.

Adherence to these principles will shape the selection of the administrative structure that is most appropriate for overseeing the financing and managing of public facilities. Where it is constitutionally permissible, and politically feasible, the administrative structure should allow the capital and operating costs of the facility to be financed out of some type of user fee. User fees are not fiscal gimmicks that circumvent local revenue or expenditure limitations. If properly designed, they can serve the same function that freely determined prices serve in private markets for private goods, which is to provide incentives that promote efficiency both in the production and consumption of goods. User fees are public prices for public goods. Prices encourage consumers to choose efficiently between alternative goods and services, and they signal to the supplier whether to expand, reduce, or maintain the existing level of output or capacity. They are, therefore, a valuable source of information to both planners and managers—a source of information that is almost impossible to replicate with techniques that estimate demand indirectly.

A State Infrastructure Bank[1]

Perhaps the most comprehensive and, potentially, the most effective way of planning, budgeting, and financing major public works projects is through a state infrastructure bank. This is still an experimental proposal, however, that will require extensive legislative action at the state and federal levels.[2] In short, the infrastructure bank is a mechanism that would pool federal infrastructure grants and state funds to capitalize a bank that would make zero- or low-interest rate loans available to local jurisdictions to finance the construction of major public facilities. The bank would operate a revolving loan fund and would also help localities finance the project by providing technical assistance to the locality, assisting in arranging lease, net lease, or service contracts with private corporations (see Chapter 5, above), and by purchasing the tax-exempt debt of the locality (acting as a bond bank).

The first proposal for a state bank emerged from New Jersey's Department of Environmental Protection, which faced requests for $2.4 billion to fund sewer upgrading from 230 local communities while federal grants from the Environmental Protection Agency (EPA) would be at most

$385 million between FY 1982-1985. Instead of providing grants to a small fraction of the eligible communities (the state estimates between 11 and 13 projects could be funded), the federal funds would be provided as loans and the repayments recycled. The concept was extended to transportation and water supply systems after an analysis of the projected needs showed that projected federal and state resources fell far short (Figure 1).

The pooling of state and federal resources and providing federal funds in a lump sum, not tied to specific projects and without other restrictive strings, would be a major step toward removing the barriers to efficient capital budgeting described in the preceding chapter. The following is a summary of New Jersey's proposed bank:

- *Structure of the Bank.* New Jersey proposes to combine currently available federal and state monies and use

Figure 1

Needs and Resources for Infrastructure in New Jersey

$ in Billions

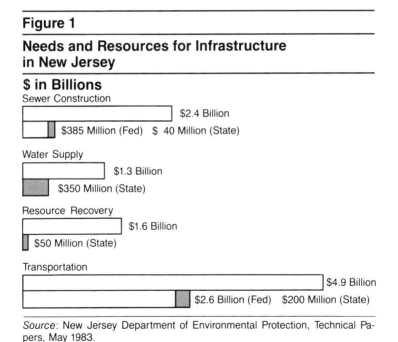

Sewer Construction
$2.4 Billion
$385 Million (Fed) $ 40 Million (State)

Water Supply
$1.3 Billion
$350 Million (State)

Resource Recovery
$1.6 Billion
$50 Million (State)

Transportation
$4.9 Billion
$2.6 Billion (Fed) $200 Million (State)

Source: New Jersey Department of Environmental Protection, Technical Papers, May 1983.

them in a revolving loan fund. These monies would be deposited in an infrastructure bank that would function much like a commercial bank. The bank's assets would be managed professionally and would earn interest, thus increasing the total money available. Customers would be state, county and local governments. As the loans are repaid, the proceeds would go back into the bank, providing capital for new projects. These loans would be financed on the local level through user fees.

- *Source of Funds.* New Jersey would capitalize the bank through state-authorized bond issue proceeds and state appropriations. In addition, the state is requesting the ability to use the funds available through the EPA Wastewater Treatment Construction Grants program. In the future, the state envisions contributions from the private sector as well.
- *Federal Role.* Under the federal Clean Water Act, funding is allocated to states to finance the construction of sewerage treatment facilities. From this allotment, local governments receive grants for specific projects based on a priority list developed by the state.

New Jersey has requested Administration support for an amendment to federal law that would allocate wastewater treatment grant funds to the state rather than to individual projects. The state would then provide loans at no interest through the bank to the individual projects. In some cases, the local share of project costs could be loaned at special low interest rates. Projects would continue to be selected through the existing process, and all current federal rules and restrictions would apply.

All funds provided by the federal government for use in the construction grants program would be segregated within the infrastructure bank. In other words, federal funds appropriated for wastewater treatment will be used only for projects eligible under the federal program.

- *Amendments.* Specifically, the amendment provides that the State of New Jersey (or any other state) can use the funds normally allocated under the Clean Water Act Wastewater Construction grants program for loans, or a combination of loans and grants;
  —that the state will be the recipient of the funds, rather than the local community; and

—that the federal government will provide the state's allocation in a lump sum, rather than on a project-by-project basis.

The overall structure of the bank and the flow of funds is shown in Figure 2. The bank offers several advantages. First, since it is designed to respond to requests from many different state agencies and local governments, the bank is in an ideal position to encourage and coordinate capital planning and budgeting procedures by these units. The project evaluation procedures used by the bank in determining whether or at what level to fund a particular project will require applicants to undertake more careful analyses. Application and evaluation procedures can also be made more compatible for different categories of the project. This is likely to provoke some resistance from local governments that are used to the application procedures associated with federal programs and have established strong connections with federal agencies. It will provoke even stronger objections from those jurisdictions that would have received grants from EPA and would receive only loans under the proposed bank. Although the transition may be difficult, the long-run improvements in the effectiveness of planning and budgeting by state and local governments will more than compensate.

Second, since the bank operates a revolving loan fund, it will be able to project available funds for a long period into the future, allowing for *long-term* capital planning by state agencies and local governments—in fact, long-term planning will be essential if the bank is to maintain a sound bond rating and earn a high rate of return on its investments.

Third, user fees will be increased to service the debt to the bank and any state or local bonds issued to finance the local share. This can improve the efficiency with which the facilities are used and managed. The increase in user fees will be much less than if the projects were financed locally. In the area of wastewater treatment, New Jersey estimates that the increase in user fees would be 33 percent when compared with a 55 percent federal grant. Total local funding would lead to an increase in user fees of 73 percent (Figure 3). Raising user fees is never easy for local

Figure 2

New Jersey Infrastructure Bank Functional Diagram and Flow of Funds

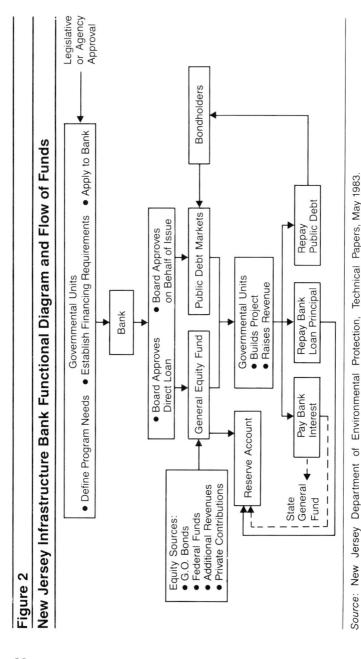

Source: New Jersey Department of Environmental Protection, Technical Papers, May 1983.

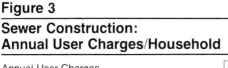

Figure 3
Sewer Construction:
Annual User Charges/Household

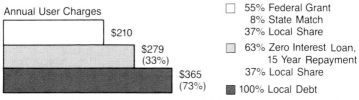

Source: New Jersey Department of Environmental Protection, Technical Papers, May 1983.

governments, which has led to less direct, and less efficient, forms of financing. The bank could provide a politically expedient mechanism for raising user fees because it assumes the responsibility for increasing user fees—a responsibility that is very difficult for local officials to assume alone.

Fourth, the bank provides an easy mechanism for coordinating planning and budgeting. The bank's resources are finite and the plans submitted by competing agencies must be reconciled with the resources available for lending or for issuing debt. Finally, the bank provides a mechanism for cutting the costs of debt financing by ensuring an orderly marketing of state and local obligations. Many states provide no marketing organization and the costs of debt issue may increase if several agencies, authorities, and localities attempt to issue debt in a short-time period. The bank could prevent this.

There are several issues, however, that need to be resolved during the policy debate precipitated by New Jersey's innovative action. First, how is the bank to be governed and managed? New Jersey envisions a privately managed bank but there has been considerable pressure from the State Legislature to make the bank accountable through the annual budget process. If the bank must seek legislative approval for each project, it becomes little more than a super-state agency, and many of the advantages outlined above will be lost. But if there is no accountability, the bank's actions will not reflect the policies of either the chief executive or the legislature. Accountability may be met through a board of trustees that would include appointees of both the governor and the legislature and some ex-officio members. Some legislative review would also be appropriate, perhaps as the responsibility of a standing committee.

Second, in some states there will be constitutional problems to the enactment of an infrastructure bank. In Kentucky, for example, the constitution mandates that revenues raised for highway projects go directly to those projects and not filtered through a state bank. In Utah, earmarking state revenues for a specific purpose is constitutionally prohibited. Other states have poor experiences with independent authorities such as a bank and might be unwilling to experiment with one. It will be difficult, although not impossible, to administer federal grants programs that flow in part directly to localities for specific projects and in part to state infrastructure banks.

Third, it will be necessary for the state banks to develop sanctions to ensure that localities comply with the basic goals of clean water, adequate transportation, and safe disposal of waste that have spurred the federal programs. EPA has experienced problems in compelling some localities to develop the required plans for secondary and tertiary treatment, in spite of their ability to withhold federal funds. State banks may find it more difficult because they are not as distant and cannot offer such generous terms.

Other states are developing their own version of this concept. Massachusetts is proposing a bank that would not intercept current programs such as the wastewater treatment grants. It will issue revenue bonds backed by user fees for a wide variety of projects and would not make subsidized loans to localities. The concept of a state infrastructure bank is one of the most promising ways of making better use of federal, state, and local resources to develop and maintain the nation's public works.

Authorities and Special Districts

The development and management of publicly owned facilities is normally conducted by a state agency, a public authority, or a special district. (Unless, of course, the facility has been turned over to a private firm for management—see Chapter 5.) Each of these mechanisms has certain advantages and disadvantages, and their applicability will depend upon the constitutional structure and traditions of each state. The activities of state or local government agencies can be coordinated more effectively than can the activities of authorities or special districts and are more responsive to the changing priorities and objectives of chief executives and legislatures. Authorities and special districts, however, are more likely to maintain facilities in better condition and less likely to make politically expedient but economically unwise cuts in maintenance budgets than are state agencies. Authorities and special districts can usually hire more competent and higher paid staff than can state agencies and can issue debt that does not require voter approval and is not subject to debt limitations.

Distinguishing between the three types of organization considered in this section is not always easy. Some state agencies have separate bonding capacity and the ability to enter into contracts that allows them to approximate public authorities. For example, the California Department of Water Resources is able to issue bonds to finance construction on the central California project backed by the revenues from water purchases by 11 local water utilities. It can act very similarly to an independent water authority.

Public Authorities

Public authorities are governmental units established by legislative action to meet some specified public purpose, such as the construction and maintenance of a bridge, convention center, airport, etc. They are allowed to issue tax-exempt bonds and are accountable to the legislature. The number and size of public authorities have grown in the past two decades. Most states have created public authorities to finance and manage ports, to issue mortgage revenue bonds, and to run publicly constructed facilities such as sports stadiums, convention centers, and turnpikes. Authorities are responsible for issuing more than half of all long-term tax-exempt bonds. In spite of the popularity of public authorities, they have become the object of a growing amount of criticism on the grounds that they are not held accountable.[3] The much-publicized bankruptcy of the Washington Public Power Supply System (WPPSS) may raise the cost of borrowing to some public authorities because it has called into question the efficiency and business expertise of this type of public agency (*Business Week*, July 11, 1983). While very few are guilty of the mismanagement that appears to have plagued WPPSS, all will be scrutinized more carefully.

The growing use of public authorities is the result of several factors. First, they allow the state much greater flexibility in issuing debt. With many states subject to limitations on the level of general obligation debt, and most states requiring voter approval for any general obligation bond issues, authorities are an effective way of financing capital projects quickly and with little effort. For example, in Michigan, the State Building Authority, created in 1976, is empowered to issue bonds to construct, acquire, improve, enlarge, and lease facilities for use by the state and institutions of higher education. The state leases the facilities from the authority and pays a rent that covers bond redemption and interest as well as estimated expenses for operation, maintenance, and repair of each facility. This authority allows the state greater flexibility in issuing debt. The state constitution requires voter approval for general obligation bonds. Special authorities are exempt from this provision because their bonds are considered revenue bonds. Studies have found that local governments that use public authorities tend to issue more total debt than those that do not (Shaul, 1980). Most authorities have established good records in the efficient management of public facilities.

Second, authorities are often exempt from the laws covering money management and contract bidding that govern state agencies. For example, the Ohio Building Authority, which borrows money from the Ohio Workmen's Compensation Fund to build facilities, issues bonds when the building is complete and services the debt from lease payments from the state government. An authority official is reported as saying that "the authority was established to avoid the red tape and controls associated with state agencies. State agencies usually cannot hold large sums of money for more than a two-year state budget cycle, but the authority can tie up large sums of money for longer periods" (U.S. GAO, November 1982, p. 35).

Third, since the authority has a specific and defined function, it gives much greater attention to maintenance than do state and local agencies. In a comprehensive study of sewer and water systems in major cities, the Urban Institute found that when these systems are "taken out of the budget process and entrusted to independent authorities or put on a full enterprise basis, the capital stock condition of these systems are superior" (Peterson and Miller, 1981). The studies attributed this to related reasons (page 41 ff):

> First, these operations are insulated from the immediate budget pressures of city government. Capital maintenance and repairs have in the past been singled out by general purpose governments for budget reductions during periods of financial pressure. . . .
>
> Second, prudent operation of a sewer and water system will accumulate depreciation reserves for future capital replacement. These cash reserves are tempting to city officials struggling to cope with short-term cash flow problems . . . . Without institutional separation between water and sewer systems on the one hand and general purpose governments on the other, recent experience proves these reserves to be vulnerable.
>
> Third, independent authorities or enterprise funds are required to run their operations on self-financing businesses. They set prices for water and sewer services at rates that recover full costs . . . water prices are significantly higher in independently operated systems than in systems that are operated as city departments, where severe underpricing is frequently a political fact of life.
>
> Fourth, independent authorities and enterprise operations are better able to gain access to capital markets.
>
> Finally, public authorities are rarely subject to state or local civil service hiring guidelines and salary levels. They are able to attract experienced and able administrators from private, for profit enterprises.

At the same time, it is important to weigh the disadvantages that arise from institutional independence of public authorities. The bankruptcies of New York's Urban Development Corporation in 1975 and the Washington Public Power Supply System in 1983 are testimony to the fact that authorities can make serious mistakes on a very large scale, and that these mistakes can go undetected for a long time. There are few market tests of the efficiency of the operations of public authorities. While legislative committees and the executives examine the procedures and policies of state agencies through the annual budget process, there is much less oversight over the activities of authorities. Anne Marie Hauck Walsh, who conducted a detailed study of public authorities during the 1970s (1978), has concluded that: "The problem is that no one pays attention to these authorities as a class in the governor's office or anywhere else. The ideal arrangement would be to have someone who knew how they worked and what they could do. Someone who could keep an eye on them" (Henriques, 1982, p. 28). A lack of accountability is a major problem that few states have dealt with successfully. Rewriting the enabling legislation of all state-chartered public authorities would be a cumbersome process. Expanding the oversight function—either through the Office of the State Comptroller, the state Public Utilities Commission or the Attorney General, or through a permanent legislative review commission (or both)—may be effective, but only if sufficient funds are appropriated to hire enough competent staff.

Another problem with public authorities is that they reduce the flexibility of state and local governments in assigning budget priorities. Neither the governor nor the legislature can balance the needs of different types of infrastructure investments if those investments are controlled by independent authorities. If changing fiscal and economic conditions demand a radical change in expenditure priorities, then a state with many public authorities will be less flexible in its response.

Finally, the independence and lack of systematic oversight enjoyed by authorities may lead to corruption and abuse. Walsh (1978) concluded, in her report of authorities based on data from the early 1970s, that there were relatively few instances of open scandal, but has recently concluded that, where local politics are corrupt public authorities "can become the ideal vehicle for carrying out that corruption" (Henriques, 1982, page 27). Without explicit oversight mechanisms, authorities may be a dangerous management structure for state or local governments to turn to.

Special Districts

A special district is essentially a limited and localized public authority. It has the ability to issue bonds for local capital improvements—such as streets, sidewalks, recreation areas, or the rehabilitation of industrial and commercial space—backed by a special assessment that is often an addition to the local property tax and may even be collected by the local government and redistributed to the special district. For example, in 1975, the Michigan legislature authorized the Detroit Downtown Development Authority as a means by which the city could designate sections of its downtown for revitalization. All increases in tax revenues from the designated areas have been put into a special fund that is to be used for land purchases, utility relocations, and the operation of the authority. No voter approval is necessary on how the tax revenues are used.

In rural areas, especially in the West, the special district has been widely used to finance the construction and maintenance of irrigation systems. Special districts are suitable where a capital project benefits a well-delineated geographic area. The major advantage to local governments is that those directly benefitting from capital improvements pay for them. Also, the local government need not use its tax base or general obligation bonds for financing highly specialized projects (Shaul, 1980, p. 14). The district structure allows local preferences for public facilities to be satisfied more effectively than through city- or county-wide projects. They also provide developers with a way of financing overall project facilities —although if the development fails, responsibility for the infrastructure may revert to the local government.

Some special districts require a vote of the majority of local property owners before the special assessment may be levied. Under New York State's City Business Improvement District Law, local businessmen can voluntarily assess a property tax surcharge providing that owners of more than 50 percent of the assessed property approve the plan and that the organization and intention of the district are approved by the state Comptroller. Management of the district includes members appointed by the city government.

These districts tend to be more accountable than public authorities and, if management and oversight are effectively structured, their activities can be coordinated within the overall capital plans of the local government. Enforcing a local vote on setting up a special assessment district ensures that its activities are compatible with local demands.

The problem with the efficient use of the special district mechanism is that it is difficult to establish. Considerable expenditures of time and money are needed before a special assessment district can be established. Since they are often most needed in low-income neighborhoods, local resources to support the organization effort may be limited. Several states have attempted to overcome this problem by providing tax credits to corporations that donate money to neighborhood-based development organizations.[4] This is intended to encourage closer ties between not-for-profit organizations and the local business community. Other states have used their Small City Block Grant monies to support these structures.

Conclusions

Different types of administrative structures have different strengths and weaknesses in managing public capital facilities. A state agency—usually headed by a gubernatorial appointee—is the most responsive to a chief executive's directives, but is also likely to have shorter time horizons than the more independent authorities and special districts. Since the latter have relatively specific purposes, however, responsibility for overall long-term capital planning and for coordinating the data collection and activities of the various entities involved in capital facilities planning and management is probably best placed in the governor's office or in a special state agency.

Under ideal conditions, most public facilities—airports, ports, water and sewer systems, convention centers, etc.—are best financed and managed through single purpose public authorities. Authorities will take a long-term view of their responsibilities, will not be able to subsidize the facility from general revenues, and if they can structure user fees effectively, will finance the construction and operation of the facility both equitably and efficiently. Authorities can span several jurisdictions and therefore capture any economies of scale. Problems have occurred, however, when the powers of public authorities have been defined too broadly and where legislative and executive oversight have been lax.

There can be no hard-and-fast rules for determining what type of public facilities are best managed by authorities, districts, or agencies. It will depend upon the strength of the planning process and the ability to hold independent agencies accountable, as well as upon the overall allocation of responsibility between the public and private sectors. A prudent and effective public investment strategy will involve a state evaluating all these types of management mechanisms. To reject the public authority because of unfortunate past experience will limit a state's ability to improve the quality of its public works investments as surely as relying on the public authority, regardless of potential abuse.

CHAPTER VII NOTES

1. This discussion draws heavily upon material prepared by the New Jersey Department of Environmental Protection and summarized by Governor Thomas H. Kean before the U.S. Senate Committee on Environment and Public Works, February 8, 1983.

2. The New Jersey proposal would require amendments to the federal Clean Water Act and to the legislation governing the use of highway grants to state and local governments.

3. The most extensive study was conducted by Walsh (1978). See also Henriques (1982).

4. The states with the most extensive programs are Pennsylvania, Wisconsin, and Florida.

Timing Public Works Investments[1]

EACH TIME THE NATIONAL ECONOMY FALTERS, Congress responds with special, countercyclical public works programs. The recent recession is no exception. The desire to create jobs led to an increase in highway user taxes to raise annual spending on highway construction and rehabilitation, and to the passage of a $4 billion jobs bill that is a pot-pourri of accelerated public works, income support for the homeless, and other emergency measures. The desire to stimulate the economy has led to many other proposals, ranging from a revived Works Progess Administration to a federal infrastructure bank. Yet past experience suggests that federal countercyclical efforts fail to meet their desired goals. They have failed to generate new jobs, have provided money too late, and have not targeted resources where they are most needed.

States could provide a more effective program to use capital investments countercyclically by setting up stabilization funds that would accumulate revenues during years when the local economy and tax revenues are growing relatively rapidly and would use the funds to finance preplanned projects when the local economy softens. These stabilization funds could be used to provide money directly to state infrastructure banks (described in the preceding Chapter). Although these funds could not help recovery from the recent recession, establishing them now will help the next time the national or local economy experiences a recession. How these funds could work is described later in this chapter.

These stabilization funds would not be a complete solution to the infrastructure problem. Rather they would provide a mechanism through which increases in expenditure made possible through improved planning and management and more effective financing techniques could be timed to help moderate the sharp and wasteful cycles of employment in the construction industry. This chapter does not advocate the funding of public works projects merely on the grounds of providing jobs during a recession.

Why Federal Countercyclical Public Works Programs Have Failed[2]

Every recession since 1960 has led to countercyclical public works programs (Vaughan, 1980). These efforts have been largely ineffective because of the following factors:

- *Policy Delays.* The onset of a recession is not immediately recognized and must be geographically widespread to be politically accepted. Political debate and administrative activity use up valuable time. Thus, it takes many months, even years, to spend countercyclical funds.
- *Displacement.* Because of the uncertainty over the policy, federal countercyclical grants to state and local governments often do little other than substitute federal for local funds with no increase in total public expenditure. The result is some fiscal relief for state and local governments, but little increase in capital spending.
- *Targeting.* Areas differ widely in their cyclical behavior. Federal targeting rarely reflects these differences, and the jobs created by federal programs are not created in those areas experiencing the greatest cyclical unemployment.
- *Job Requirements.* Construction jobs are of very short duration and require skills that few of the unemployed possess.

Lags in Allocating Money

To be effective, a countercyclical program must be timed correctly. Federal stabilization programs have been ineffective because they are undertaken too late. Expenditures should increase as soon as possible after a cyclical downturn and should decline as soon as possible after

109

recovery begins. The record at the federal level has been poor. Money provided under discretionary countercyclical programs has not been spent until between two and four years after the cycle downturn.

Delays occur at four stages: (1) recognizing that a cyclical downturn has occurred, (2) passing the appropriate legislation, (3) taking the appropriate administrative action once the legislation has passed, and (4) enrolling the program participants. Most money has not been spent until four years after the economy has entered a recession (Vaughan, 1980).

Nothing can be done to avoid delays in recognizing a recession, although they will be shorter at the state level than at the national level because, at the state level, there is less of an aggregation problem than at the national level. A lead state will have experienced unambiguous indications of a recession long before the nation as a whole.

To avoid local legislative delays, the program should be designed to function as automatically as possible. The countercyclical fund described in the following section could undertake expenditures automatically when target economic indicators cross threshold levels.

To avoid administrative delays, states should work with local governments or through infrastructure banks to ensure that there are suitable projects "on the shelf," ready for immediate ground-breaking during a recession. Planning a public works project takes time—it must be coordinated with local economic development strategies, meet environmental and affirmative action requirements, and acquire the necessary zoning actions and other regulatory permits in advance.

Substitution and Displacement

The ability of the federal government to stimulate economic activity by increasing expenditures during a recession is impeded by substitution (i.e., the crowding out of private sector investment by federal borrowing to finance countercyclical spending) and by displacement (i.e., reductions in state and local financing in response to increases in federal grants). Both substitution and displacement mean that a job created by federally financed programs does not represent a net increase of one job in the economy, since private sector and local public sector employment is reduced.

Displacement is a major obstacle to the implementation of federal countercyclical public works programs that are channeled through state and local governments. Total public sector spending increases by much less than the value of the federal grant. Estimates of displacement suggest that displacement within one quarter of receiving federal funds is between 0 and 30 percent and may approach 100 percent after two years (OMB, 1979). State and local governments cannot be blamed for the problem of displacement. No regular countercyclical program has been set up. When the federal government reacts to a recession and hastily allocates funds for public works or public employment, there is little that can be done at the state or local level within the federal time constraints other than to undertake projects that were already on the drawing board, or to hire employees for slots that already existed. This means that there is little increase in public capital expenditures. Expenditures are not rescheduled from expansionary years to recession periods. The state stabilization fund strategy described below would not involve any increase in spending; rather, it would reschedule spending from periods of rapid growth to periods of slow growth.

The countercyclical Local Public Works (LPW) program probably suffered from close to a 100 percent displacement rate. That is, in spite of allocating $6 billion in federal funds, there was virtually no increase in state-local construction expenditure (Gramlich, 1978; OMB, 1979). From the first quarter of 1976 to the first quarter of 1977, total state and local construction expenditures fell by $7.7 billion in current dollars. At the same time, federal capital grants for highways and sewers rose by over a billion dollars, real income grew by 4.5 percent, and state and local bond rates fell by almost 1 full percentage point. This puzzling decline of 27 percent in real terms can be attributed to the passage of the Local Public Works Act in August 1976, after nearly a year of protracted deliberation in Congress. Local officials had ample notice of forthcoming federal funds and, therefore, an incentive to delay their own construction in the hope that the federal government would pay for some of their projects. Until regulations were issued and the allocation formula determined, officials did not know which projects would be funded. The Economic Development Administration received $22 billion worth of applications for the initial $2 billion allocation. Many projects would have been built regardless. Rather than stimulating the economy, LPW actually delayed the economic recovery during late 1976 and early 1977.

Displacement can be avoided in a state-based program by setting up a permanent countercyclical program that identifies, in advance, specific projects and activities that will be undertaken when the local economy slows down. If specific projects are identified in advance by state governments, there is still displacement over time, but there is an increase in public spending when it is needed—during a recession. The purpose is not to increase aggregate spending over the course of the cycle, but rather to shift expenditure from expansionary to recessionary years. Ongoing stabilization programs—which federal efforts are not—would reduce displacement. And, specific project identification in advance, would eliminate it.

Targeting

The third reason why federal countercylical programs have failed is that they have not been targeted in a way that reflects local cyclical behavior. This failure arises from the following factors:

- There is a wide variation in cyclical behavior among states, and even among areas within a state. Since federal countercyclical programs are triggered by national indicators, aid is often inappropriately timed for local cyclical problems.
- The process of creating a congressional consensus leads to allocation formulas that do not reflect local needs.
- Many of the jobs created by the spending of funds in a given area spill over across geographical boundaries and, therefore, do not help the targeted local economy.

Many of the problems of targeting will exist at the state level as they do at the federal level. The political machinations necessary to achieve consensus can just as easily lead to ineffective and inequitable allocation formulas. A state-based program will only be successful if it is designed and operated with the full cooperation of, and contributions from, local jurisdictions. Encouraging localities to buy into the stabilization funds provides some measure of local control over allocations. Allowing local participation in the program design and implementation would improve allocation procedures. Federal strings may be necessary, however, to ensure an equitable allocation of funds among jurisdictions and to ensure that the programs reach out effectively to those really in need.

Job Requirements

The type of jobs generated on public works projects are not accessible to many of the hard-core unemployed. If the projects are undertaken when the economy is in the depths of the recession, they will draw upon the unemployed construction workers. But if they are delayed until the economy recovers, the added funds are likely to contribute to inflating construction costs.

A 1979 Office of Budget and Management (OMB) study found that only about one fourth of those hired under the LPW program were unemployed on the day before hiring. Two-thirds of all the jobs created required specialized construction skills. Most of the unskilled jobs lasted less than one month.

A State Countercyclical Strategy[3]

A state's own economic stabilization strategy is certain to be more effective than federal efforts in softening the blows of recessions. It will not wipe out recessions; indeed, it will only make a modest dent in the economic problems that a recession causes. But it will lead to a more rational deployment of countercyclical resources and provide more targeted assistance to those affected by an economic slowdown. The strategy should be based on a stabilization fund—following the Michigan model—which is built up during periods of relatively rapid growth and spent during recessions. The fund would be used to finance countercyclical programs that would include public works,

113

public employment and training, and intrastate antireces-sionary fiscal assistance. These programs would function automatically (subject to legislative review), with funds released when local economic indicators crossed predeter-mined thresholds. Each program would be keyed to a different indicator since each program addresses a different countercyclical goal.

The stabilization fund offers state governments many advantages:

- *It requires no increase in state and local spending.* Through the fund, state and local spending is resched-uled and redistributed more evenly over the course of the cycle. Public works are concentrated in times of slack demand for construction activity, and the high level of transfer payments that a recession entails are paid for, in part, during boom years. Expenditures, therefore, re-flect average revenues over the cycle rather than the present year-to-year budgeting that has led to such a switchback in public spending.

- *It maintains the integrity of appropriations during re-cessions.* All too often, recession-induced shortfalls in revenue necessitate the cutting back (or even the cutting out) of programs, especially infrastructure maintenance, for which the legislature has appropriated funds. The sta-bilization fund avoids the need for such wasteful surgery.

- *It reduces the temptation during good years to expand public programs beyond a sustainable level or to tem-porarily cut taxes.* Surging revenues during economic booms, when social services are at a low level, often en-courage the expansion of existing programs or the addi-tion of new programs that would not be undertaken if the budget constraint were tighter. A state surplus fre-quently leads to political pressure to cut taxes. From the perspective of the overall cycle, these surpluses are not real. They are matched by potential deficits during slumps. The stabilization fund avoids the appearance of such surpluses.

- *It may actually reduce spending in the long run.* By encouraging a longer term approach to budgeting, re-ducing the full impact of recessions, and retiring capital expenditures, the fund may actually help reduce state and local expenditures.

A State Stabilization Fund

To finance the necessary countercyclical programs, and to reduce excessive growth in spending during periods of rapid expansion and inflation, states must set up stabilization funds. Contributions would be accumulated while the economy, and state and local revenues, grow at above average rates and would be withdrawn and spent when growth and revenues falter. The fund would not necessarily be reduced to zero during every recession. Some recessions are deeper than others. By maintaining some reserve during a shallow decline, the fund would be larger when a deep recession threatens.

Establishing a dedicated fund is essential. States will not set aside surpluses for a rainy day without a special program. The natural tendency to run surpluses during good years generates political pressure either to cut taxes or to start new programs, leaving little fiscal flexibility to face the ensuing recession. While the taxpayers' revolt maintains a head of steam, this pressure is unlikely to abate. In 1979, as the nation's economy neared recession, many states undertook wholesale tax reductions. Property taxes were reduced in 22 states, assessments were curbed in four, personal income taxes were cut in 18 states, and sales taxes were cut in 15. The principal beneficiaries of these tax cuts will be the relatively affluent. The equity and efficiency of the state-local tax system have been reduced, and as the recession makes inroads into revenues, these same states will demand federal assistance. In fact, experience of past booms and recessions suggests that the surplus is not really a surplus at all. The gap between spending and revenues should not be judged on a year-to-year basis but over the full cycle. Enough of the expansion years' bonuses should be set aside to allow for the continued operation of state and local services and to fight recession during the bad years.

Stabilization funds could draw upon four sources of money:

1. *Federal Matching Funds.* Instead of waiting until

late into a recession to provide federal grants these matching monies should be made to stabilization funds annually, allocated according to each state's own contribution, with allowance for local fiscal conditions.

2. *State Tax Revenues.* These revenues would include the cyclical component of a volatile tax such as the personal income tax.

3. *State Borrowing.* This source includes bond issues for designated countercyclical projects, in those states in which this is permitted.

4. *Contributions from Local Jurisdictions.* These include payments by local jurisdictions into a "recession insurance fund," from which they could draw as revenues fell.

How big should a state stabilization fund be, and at what rate should states contribute during good years? Ultimately, the best answer to these questions will be provided by experience accumulated through the operation of stabilization funds. The only experience thus far has been with the Michigan stabilization fund and that accumulated $300 million in the two years before the State's economy entered the present recession in the fall of 1979 from which it has yet to recover.

Calculations presented in a Council of State Planning Agencies (CSPA) report published in 1980 suggest that it would be possible to have reserves of nearly $20 billion available for countercyclical public works with modest contributions from federal, state, and local sources during expansionary periods.

States would determine how much they would pay in during favorable economic times based upon the volatility of their economies and revenues. These state payments could be matched by federal grants (see Vaughan, 1980). States should adopt an automatic triggering mechanism in which both contributions and expenditures are triggered by the performance of the state economy relative to threshold indicators. An automatic procedure leads to a more rapid response than a discretionary procedure and lessens the temptation to indulge in short-run tax cutting or spending increases rather than accumulating a necessary surplus. The Michigan fund (see Table 3) determines

TABLE 3

Michigan State Stabilization Fund

As enacted, the countercyclical budget and economic stabilization fund is designed to attack the two problems of cyclically low revenues and high unemployment. The law establishes formulas by which money is deposited in the fund and by which withdrawals can be made. Major provisions are as follows:

Budget Stabilization

1. All transfers into or out of the fund will be based upon the annual growth of adjusted Michigan personal income (MPI) in the current calendar year.

2. Adjusted Michigan personal income is defined to mean total state personal income minus transfer payments (nontaxable income received from the government) deflated by the Detroit Consumer Price Index so as to remove any inflationary bias. Transfer payments are deducted so that the full impact of the cycle is identified.

3. When the adjusted MPI grows by more than the pay-in trigger level of 2 percent, the percentage excess will be multiplied by the total general fund/general purpose (GF/GP) revenue accruing to the current fiscal year to determine the amount to be transferred from the general fund to the stabilization fund in the coming fiscal year.

4. When the annual change in adjusted MPI is less than the pay-out trigger level of zero percent, the percentage deficiency will be multiplied by the total GF/GP revenue accruing to the current fiscal year to determine the amount to be transferred from the stabilization fund to the general fund in the current fiscal year.

Examples: If GF/GP revenue is assumed to be $3 billion and the adjusted MPI change from the prior year is assumed to be: Case 1: +7 percent; Case 2: +1.5 percent; Case 3: -4 percent, application of the formulas would be:

Case 1: 0.7 - .02 = .05 × $3 billion = $150 million pay-in to fund next Fiscal Year (FY),

Case 2: .015 is between .000 and .02 = no pay-in or withdrawal.

Case 3: -.04 × $3 billion = $120 million withdrawal during current FY.

It was not intended that the budget stabilization fund would entirely eliminate the problems posed by revenue fluctuations. Its purpose is to ameliorate the problem by reducing the extreme peaks and valleys.

Economic Stabilization

1. In any quarter following a quarter when unemployment averages 8 percent or more, the law provides that an amount may be appropriated from the fund for countercyclical policy as shown below:

Percent Unemployed in Most Recent Quarter	Percent of Fund Available for Economic Stabilization During the Following Quarter
8.0 – 11.9 percent	2.5 percent
12.0 percent and over	5.0 percent

Example: If the stabilization fund balance is assumed to be $200 million and the rate of unemployment is 9 percent for the quarter ending March 31,

---

TABLE 3 (Cont.)

1979, the fund could be used as follows in the April-June quarter: .025 × $200 million = $5 million for countercyclical programs.

    2. The funds appropriated for economic stabilization may be used for capital outlay, public works and public service jobs, refundable investment or employment tax credits against state business taxes for new outlays and hiring in Michigan, or any other purpose the legislature may designate by law which provides employment opportunities counter to the state's economic cycle. Obviously, the latter purpose is subject to very broad interpretation.

---

contributions based upon the growth in real personal income, and triggers the release of funds for budget balancing (antirecessionary fiscal assistance) based upon real income growth and for economic stabilization (public works and public employment) based upon the state unemployment rate.

In brief, the law states that payments will be made into the fund when the adjusted Michigan personal income (MPI) annual growth rate exceeds 2 percent, and, withdrawals from the fund may be made in four situations: (1) when the real MPI decreases, (2) when quarterly unemployment exceeds 8 percent, (3) when revenue falls short of statutory estimates (without change in the tax rate or base), and (4) in an emergency upon two-thirds vote by each house.

Unfortunately, the Michigan fund had been in effect for less than two years when the severe and prolonged recession began in 1979 and from which the state is only now beginning to recover. The fund had accumulated only about $300 million, but it serves as a model that could be followed by all states with the assistance of the federal government.

The federal government could allocate the $3 to $4 billion a year it has spent on discretionary countercyclical programs among state stabilization funds according to the states' own contributions in the form of a matching grant, with some allowance for differences in fiscal capacity (see Vaughan 1980). During the early years of this policy, when only a few states have such funds, it would be necessary to fix a matching rate. When almost all states participate, the total $4 billion would be divided among the states according to how much they have contributed, leading to a lower matching rate when state contributions are large, and a high rate when they are small. The on-off approach to countercyclical policy, which changes from one recession to the next, has stunted the development of local stabilization capacity. It has also meant that federal expenditures have little stimulative effect because they tend to displace state and local funds. The federal and state contributions could be augmented by contributions from county and city governments.

Countercyclical Activities

The stabilization fund could disburse revenues to fund public works projects, to provide countercyclical fiscal relief to localities, and to assist the unemployed to pay for training. The major function of public works projects is to develop economic and social infrastructure. The public works component of the countercyclical strategy is targeted at providing employment for idle resources in the construction and building supplies industries. Special construction projects should not be undertaken solely to provide jobs for the unskilled or inexperienced. There are two reasons for this: (1) the average duration of low-skilled jobs on public construction projects has only been about 20 days, and (2) because of skill requirements and the nature of the construction labor market, few jobs for the unskilled are generated onsite.[1]

A project for which the finances had already been raised would be started when employment in the local (county or city) construction sector shows a cyclical downturn. Because monthly local employment and unemployment data are not always reliable, there is something of a trade-off between waiting for sufficient monthly data to be sure that a downturn has occurred, and creating jobs as quickly as possible. Employment and unemployment data, while providing confusing signals, are, along with housing permits, the only readily available and timely data at the local level. States will have to experiment with developing efficient leading indicators of construction downturns.

Public works projects undertaken as part of state stabilization policy should be those that are compatible with the area's overall capital investment plan but which can be safely defined for one or two years. These may include state prisons, new roads or major rehabilitation of existing roads or highways, new recreation facilities, or major port rehabilitation. There is little that can be gained in selecting special labor-intensive projects since these are unlikely to provide the unskilled worker with more than two to three weeks work, which does next-to-nothing for his skills, work experience or income maintenance. Make-work construction projects leave the local taxpayer with bond-servicing costs and little of permanent value by way of reward.

Heavy construction projects, such as flood control, dams, levees, water supply facilities and water treatment plants, are rarely suitable for countercyclical purposes because the heavy construction industry does not behave cyclically. Moreover, the equipment and skills needed on such projects are highly specialized and may have to be imported from other areas and states, which will reduce their local job-generating ability.

For timely implementation, the planning for counter-cyclical projects must have been completed in advance, so that time-consuming public hearings and lawsuits do not delay getting construction firms to work on the projects. Preselection is obviously a difficult issue. If the project is essential to the delivery of public services, then delay may be costly and disruptive. On the other hand, construction projects should never be undertaken simply in order to create jobs. Developing the planning capacity at the state and local levels to set aside capital projects for counter-cyclical implementation will take time and will involve a new dimension to capital planning and budgeting (see Chapter 5).

Conclusions

Federal countercyclical public works programs have failed. Money has not been allocated rapidly enough to fund projects while the economy is still in a recession. Most federal grants simply substitute federal for local money, leading to little net increase in public works spending. Few of the jobs on federally funded projects have aided those most harmed by a recession.

A more promising approach would be to set up state stabilization funds that would accumulate money during periods of growth. The money would be used to pay for pre-planned public works projects as soon as the local economy softened.

CHAPTER VIII NOTES

1. This chapter is adapted from the author's "Inflation and Unemployment: Surviving the 1980s" published by the Council of State Planning Agencies in 1980.

2. For a more detailed evaluation of the failure of countercyclical programs, see Vaughan and Vernez, 1978.

3. This initiative is described in greater detail in Vaughan, 1980.

CHAPTER IX

Summary and Recommendations

DURING THE NEXT FEW YEARS state and local governments must increase the level of investments in public works and also dramatically improve the quality of those investments. Taxpayers are unlikely to support increases in taxes and user fees if they feel that the additional capital investments they are supporting are not well planned, if the facilities are not well managed, and if they do not provide tangible local benefits. Investors will not purchase bonds unless the borrower has established a record of fiscal prudence and good management.

At present, there are few state and local governments that have set up the planning, budgeting, and managing procedures that would permit them to spend more money on capital investments wisely. While all have a long list of projects that they would like to see funded, few have performed the cost-benefit analyses that allow for an informed assignment of priorities. A public capital investment program will only prove effective if it is based upon a sound assessment of what needs to be done and how best to do it. This book has reviewed some of the issues relevant to capital planning, budgeting, and managing issues. The companion volume *Financing Public Works in the 1980s* discusses how the necessary fiscal resources can be raised.

A state capital plan is necessary because of the wide range of investments that state and local governments must undertake to support industrial and residential development and the relatively limited fiscal resources that are available to meet demands for public support. Public demands for and the technology of public service delivery are constantly changing. The future is not a simple extrapolation of the past and, without the capacity to conduct strategic capital planning, state governments are unable to respond to these changes. Money is wasted in the construction of unneeded facilities, or development is delayed by inadequate water, waste disposal, or transportation facilities.

Capital Planning

Capital planning is forward looking. It is an attempt to identify the projects that will be needed to meet future demands in time for the investments to be made efficiently. Since the future cannot be predicted with certainty, planning involves contingencies such as establishing the economic and demographic indicators that will be watched in order to determine whether the underlying assumptions of the capital plan are valid, and what to do if they are not.

In spite of the growing popularity of complex econometric models to forecast the future, planning should include a strong qualitative element—mainly, attempts to predict changes in the organization and distribution of economic and residential activity that will change the type of public capital investments that will be required. The exercise of predicting the impact of technological change, changing world trade patterns, demographic shifts, and natural resource issues, for examples, is one way of imbuing those responsible for planning with a future orientation. It can also dislodge from the planning process the mystique of quantitative analysis, econometric modelling, and computer printouts, which have contributed to the separation of planning from decisionmaking.

The Politics of Planning

The purpose of capital planning is to provide decisionmakers with the information that allows them to make prudent decisions. Capital planning is not in itself the act of making decisions. Those responsible for preparing the capital plan must use some judgment in determining what issues to examine and in identifying what options to evaluate, but their function is to inform, not to select. And they can best inform by rigorous application of cost-benefit analysis to the evaluation of proposed investments. Although analytic rigor must, inevitably, bow to political considerations at times, it cannot be bent too often without the planning function losing credibility.

Capital planning cannot be effective without the full support and participation of the governor. Planners must be clearly linked to the decisionmaking process and not be preoccupied with fulfilling requirements for federal funds. Too often, the planning process has been used to justify investment decisions that have already been taken, not to explore options that are available.

What Is Capital Planning?

Establishing a planning process is not a commitment to a plan but a commitment to anticipating the future and developing a strategy to deal with future events—that is, a strategy that must continually be adapted to economic and demographic changes. Four characteristics of the planning function should shape the way that a state plans its capital investments:

- First, planning is a continuous process.
- Second, the management of the incoming and outgoing flow of information is a central element of the planning process.
- Third, planning is an alchemist's amalgam of science and art.
- Finally, setting up a strategic planning function will not be popular initially, but will gain acceptance if it is used in the decisionmaking process.

Capital planning is too important to be left to professional planners alone. It should be the top priority of the governor and should involve his direct staff and a range of managers with line responsibility from throughout state government.

The Economics of Planning

At the core of the planning process is the evaluation of alternative capital investments and programs. Rational decisionmaking requires standards for evaluation and a set of directions for applying them. The economic way of thinking provides a reasonable and logical framework for comparing complex policy choices. These standards must be applied within the institutional, legal, and political constraints prevailing in the state. They do provide the basis for high-quality and relatively objective evaluations. While they do not allow the weighing of redistributional issues, they do allow the attachment of dollar values on the likely outcomes resulting from a proposed investment.

The rigorous application of cost-benefit analysis does not require complex models. Indeed, basic economic common sense is often sacrificed when technically complicated procedures are followed.

Estimating the net present value of a proposed project—that is, the difference between the present value of the probable benefits net of the present value of the costs—allows the comparison among projects and the development of an effective capital investment plan.

It is important, however, to realize the limits of cost-benefit analysis. The future cannot be predicted with certainty—no matter how costly the computer program or the consultant used to predict the costs and benefits. There are techniques that can be employed to include risk in these estimates and to assess the sensitivity of the estimates to changes in key parameters.

Assigning Responsibilities: Public or Private

The need for public capital investments is strongly influenced by how the state and local governments allocate responsibilities for financing and administering facilities between the public and the private sectors. Some states have greatly expanded their involvement in financing what has traditionally been the responsibility of private companies, usually with the intention of attracting jobs and investment to the state. Tight budgets and declining federal grants will necessitate a reassessment of how much private investment can be undertaken by state government. States will have to establish guidelines for shifting some of the financal and managerial responsibility to the private sector. Large-scale energy developments in western states have shown that private developers are willing—in some cases, even eager—to shoulder the financial responsibility for building roads, utilities and even schools. Increased reliance on the private provision of public services and facilities—through franchising, contracting out, leasing, and even outright selling of public facilities—is being considered by a growing number of jurisdictions. Privatization should proceed cautiously. Responsibility for service delivery should only be turned over to private suppliers if the service or facility is potentially competitive enough to yield an efficient pricing system.

Capital Budgeting

Capital budgeting is the annual control device used to carry out the actions developed by the capital plan. Few states integrate their plans with annual budgets. Closer integration can be achieved through four steps:

- First, distinguish capital from operating expenditures.
- Second, assess the condition of existing capital facilities and the rate at which they are deteriorating.
- Third, estimate the impact of today's capital expenditures on tomorrow's operating budget. The capital budget should also contain estimates of the rate of depreciation of capital assets and the potential consequences on future operating budgets of increasing or reducing today's maintenance and repair expenditures.
- Fourth, alter the way the federal government provides capital grants so that states and localities can prepare rational capital budgets.

These steps do not guarantee that capital plans will be translated into capital budgets. That will depend upon the willingness of governors and their staffs to enforce the consideration of capital plans as the annual executive budget is developed.

Selecting the Management Mechanism

The type of agency that administers the management of the public facility will shape its use and how it is operated and maintained. An independent public authority usually has a revenue source that is not dependent upon annual appropriations by the state legislature and will not be subject to sudden cuts because of a general revenue shortfall. Since the authority has specific functions prescribed in its enabling legislation, it will usually maintain its capital facilities more effectively than a unit of general government. Water and sewer systems maintained by single purpose public authorities have been maintained in much better condition and at little extra cost than those maintained by local governments. Public authorities are also much more flexible than general government units with respect to employment, procurement, and other activities.

Public authorities, however, must be closely monitored. Their independence may be a source of fiscal problems as well as a managerial advantage, as attested by the bankruptcy of the New York Urban Development Corporation in 1975 and by the recent problems of the Washington Public Power Supply System. Few states have devised foolproof ways of ensuring full accountability of public authorities.

Special assessment districts offer an effective way of financing special infrastructure projects that will increase local property values and whose benefits are localized.

The most promising structure for planning and financing public works projects promises to be the state "infrastructure bank." Although no state has yet established a working infrastructure bank, several have developed quite advanced plans—including New Jersey, Utah, and Massachusetts. In essence, these banks act as bond banks (see Financing, Chapter 5) but with strong planning and technical assistance included. New Jersey has proposed to include a revolving loan fund to lend to localities capitalized from state general obligation bond revenues and from federal grants. There are several bills before Congress to provide federal grants or loans to state banks. The advantage of the state bank is that it would be a single agency to plan and finance state and local public works and would avoid the fragmentation that characterizes present efforts.

Timing Public Works Spending

Public works have traditionally been used to attempt to create jobs during recessions. Federal, countercyclical public works programs have failed, however, because they have provided funds too late, have not targeted on those areas where jobs are most needed and have largely substituted for state spending.

While a project should never be undertaken simply to create jobs, states could establish "rainy day" funds to finance needed projects during recessions. The funds could be administered through a state infrastructure bank, that would accumulate money during periods of economic growth. When the local economy turned down, the funds would be used to finance needed projects that have been temporarily deferred.

The infrastructure crisis has been popularly depicted as a lack of money. But underinvestment is only a part of the problem, and so increased spending will only be part of the solution. Poor planning, inadequate articulation of planning and decisionmaking, superficial capital budgets, and deficient management systems are, unfortunately, the rule not the exception in state government. These problems must be remedied. The necessary institutional change will not be easy or swift. The continued intervention of the chief executive will be essential. But with gubernatorial leadership, the state can make wiser use of its fiscal resources. Capital planning can save more dollars than the most elaborate creative financing mechanism. Capital budgeting can deplete those unwanted and unnecessary public works programs that have been all too often characteristic of state governments, while good management can preserve the existing capital stock.

Annotated Bibliography

The following annotated bibliography is intended as a brief guide to the overall bibliography. It classifies references under the major topics covered in this book. For overall statistical data on public investments the most useful sources are the *Survey of Government Finances*, annually; various articles in the *Survey of Current Business* (monthly, published by the Bureau of Economic Analysis); the Office of Management and Budget; and various reports by the Bureau of Economic Analysis. There are also reports published by various interest groups including the American Public Works Association, Federal Highway Users Association, American Public Transit Association, and Association of Metropolitan Sewage Agencies.

The best single source on public capital finance issues is the Municipal Finance Officers' Association, which has an extensive set of publications. The National Conference of State Legislatures, the Council of State Planning Agencies, and the Urban Institute also have entensive sets of publications and reports.

Bond Banks:
Forbes and Renshaw, 1972; Jarrett and Hicks, 1977; Katzman, June 1980 and March 1980; Solano and Hoffman, 1982.

Bonds: *(see also Bonds Banks, Credit Ratings, Debt, Housing Bonds, Industrial Development Bonds, Taxable Bonds)*
For an overview of municipal bonds and definitions and descriptions of the bond market see ACIR, 1976; Amdursky, 1981; Lamb and Rappaport, 1980; Peterson, J., 1976; Public Securities Association, 1981. See also Durst, 1981; Fischer et al., 1980; Klapper, 1980; Robinson, 1981; and White, 1979. Basic data on bonds and bond markets is reported in the *Daily Bond Buyer* and in *Moody's Bond Survey* (weekly) and is summarized in the annual reports of the Public Securities Association.

For an analysis of the behavior of the bond market, including the effects of federal tax changes, see Browne and Syron, 1979; Forbes and Peterson, 1975; Hendershott and Koch, 1977;

Herships and Karvelis, 1981; Kaufman, 1981; Kimball, 1977a; Kopcke and Kimball, 1979; Mumy, 1978; Peterson, J., 1982; Twentieth Century Fund, 1976; Viscount, 1982; and U.S. Congress, Joint Economic Committee, 1981.

Insurance: Miralia, 1980.

Marketing: MFOA, 1976.

New Jersey Qualified Bond Program: Jones, 1978; Peterson and Miller, 1981.

Small Denominations: Lehan, 1980.

Underwriting: Cagan, 1978; Silber, 1980.

Boom Towns: *(see also Energy Development)*
Brookshire and D'Arge, 1980; Gilmore and Duff, 1975; O'Hare, 1977; O'Hare and Sanderson, 1978; U.S. Department of Defense, Office of Economic Adjustment, 1981a and 1981b.

Capital Budgeting: *(see also Debt, Management and Planning)*
For a discussion of the process and benefits of capital budgeting, see Choate and Walter, 1981; Devoy and Wise, 1979; Douglas, 1977; Fujardo, 1976; Howard, 1973; Matson, 1976; U.S. GAO, 1982, 1981, and 1980; Wacht, 1980; and White, 1978. For examples of state initiatives, see Alexander, 1980; American Public Works Association, 1979; Maryland, 1980; and Rumowicz, 1980; Watson, 1983.

Condition of Public Works: *(see also Financing Public Investment)*
For a brief account of the problem, see Choate and Walter, 1981, and *Newsweek 8/2/82.* For more detailed discussions of the extent of the problem, see Abt Associates, 1980; CONSAD, 1980; Dossani and Steger, 1980; U.S. Department of Commerce, 1980. The study of the capital stock in 12 cities conducted by the Urban Institute is exhaustive; see Grossman, 1979; Humphrey and Wilson, 1980; Humphrey et al., 1979; Peterson and Miller, 1981; Wilson, 1980. See also American Public Works Association, 1981 and 1976; Beals, 1981; Buckley, 1982; Choate, 1982; Finck and Pike, 1981; Godwin and Peterson, 1982; Hatry, 1980; Lindsay, 1979; O'Day and Newmann, 1983; MacDonald, ed., 1982; National League of Cities, 1983; Watson, 1983.

Countercyclical Public Works:
See CBO, 1978; Kaus, 1982; Vaughan, 1980a, 1980b, and 1976; and Vernez and Vaughan, 1978.

Credit Rating: *(see also Bonds, Debt, Fiscal Capacity)*
The most comprehensive guides are Ingram and Copeland, 1982; Peterson, J., 1974; Smith, 1979. See also American

Bankers Association, 1968; Aronson and Marsden, 1980; Boyett and Giroux, 1978; Osteryoung, 1978; Reilly, 1967.

Debt: *(see also Bonds, Credit Ratings, Fiscal Capacity)*
ACIR, 1976; Gold, 1981; Irwin, 1979.
Danger Signals: Aronson, 1976; Groves, 1980.
Limitations: ACIR, 1980; Baer, 1981.
Management: Brown et al., 1978; Moak, 1970; Peterson, J., 1979a, 1978; Small Cities Financial Management Project, 1978; Steiss, 1975. State Assistance to Localities: Alaska Legislative Affairs Agency, 1967; Forbes and Peterson, 1978; Glaser, 1978; Peterson, 1977 (see also Bond Banks).

Energy Development: *(see also Boom Towns)*
Foster, 1977; Kolb, 1982; Leistritz and Murdock, 1981; Lu, 1977; Monaco, 1977; Rocha, 1982; Sanderson, 1977; Schnell and Krannich, 1977; Susskind and O'Hare, 1977; and West, 1977.

Federal Government: *(see also Bonds, Intergovernmental Relations)*
Grants: Executive Office of the President, 1980 and 1978; National Governors' Association, 1977 (see also Intergovernmental Relations).
And Local Fiscal Conditions: Herships and Karvelis, 1981 (see also Fiscal Capacity).
Impact on Infrastructure: Gramlich and Galper, 1973; Hatry, 1980.

Financing Public Investment: *(see also Capital Budgeting, Condition of Public Works, Lease Financing, User Fees)*
For summaries of policy alternatives, see Pagano and Moore, 1980; Peterson, n.d.; Peterson and Miller, 1981; Stanfield, 1980; Watson, 1982. For bibliography, see Buss, 1981. See also Buckley, 1982; Getzels and Thurow, 1980; McWatters, 1979; Saffran, 1979; Shaul, 1981; Wolman and Reigeluth, 1980.

Fiscal Capacity: *(see also Bonds, Debt, Intergovernmental Relations)*
ACIR, 1971; Crinder, 1978; Groves, 1980; Herships and Karvelis, 1981; Howell and Stamm, 1979; Hubbell, 1979; Matz, 1980; Rothschild, Unterberg, Towbin, 1982; U.S. Joint Economic Committee, 1981; and Wolman and Davis, 1981.

Forecasting:
Armstrong, 1978; Ascher, 1978.

Highways: *(see also User Fees)*
For a discussion of financing, see Congressional Budget Of-

fice, 1982; Langton, 1981; "Forty States Eye Motor Fuel Tax Boost," 1981.

Housing Bonds: *(see also Bonds)*
Bates and Wolfson, 1981; Harrington, 1979; Levatino-Donoghue, 1979; National Conference of State Legislatures, 1980; Peterson, G., 1979; Worsham, 1980.

Industrial Development Bonds:
Edmonds and Hoyd, 1981.

Intergovermental Relations:
For an overview, see ACIR, 1980, 1979, and 1978. See also CBO, 1978, and Zimmerman, 1976.

Lease Financing: *(see also Financing, Public Investment)*
Barnes, 1981; Dyl and Joehnk, 1978; Lubick and Galper, 1982; Mentz et al., 1980; Schellenbach and Weber, 1978; Shubnell and Cobb, 1982.

Management and Planning: *(see also Capital Budgeting)*
Amara, 1979; American Society of Planning Officials, 1977; Anthony and Hertzlinger, 1980; Bologna, 1980; Corr. 1980; Donnelly, 1980; Garfield Schwartz Associates, 1982; Hall, 1979; Hatry, 1980; Higgins, 1978; Holloway and King, 1977; International City Managers Association, 1981; Korbitz, 1976; Kunde and Berry, 1982; Lorange and Vancil, 1982; Naylor, 1979; Naylor and Neva, 1980 and 1979; Schmidt, 1979; Schneider and Swinton, 1979; Sheeran, 1976; Shepard and Goddard, n.d.; and Steiss, 1975; U.S. GAO, Nov 1982; Walter, 1980.

Public Authorities:
Beyer, 1972; Edelman, 1976; Henriques, 1982; Holland, 1972; and, for the most detailed analysis, Walsh, 1976.

Special Assessments:
Shoup, 1980.

Taxable Bond Option:
Kimball, 1978; Kopcke and Kimball, 1979; Mussa and Kormendi, 1979.

Taxes:
And Economic Development: Kieschnick, 1981, and Vaughan, 1979.
Of Mineral Resources: Zeller, 1982.
Limitations on Revenues: Bacon, 1981; Baer, 1981; Benson, 1980; McWatters, 1979; Pascal, 1980; Saffran, 1979.

User Fees:

The best overview of the principles and applications of user fees is in Mushkin, 1972. See also ACIR, 1974; Buss, 1981; Downing, 1980; Feldstein, 1972; Galambos and Schreiber, 1978; Gold, 1979; Mick, 1981; Mushkin, 1979 and 1977; Mushkin and Vehorn, 1977; Stanfield, 1980.

Airports: Feldman, 1967, Littlefield and Thompsen, 1977.

Health: Badgley and Smith, 1979, Berki in Mushkin, ed., 1977.

Highways: Abouchar, 1974; Feuer, 1978; Henion and Ford, 1981; Higgins, 1979; Langton, 1981; Smith, 1980.

Recreation: Artz and Bermond, 1970; Economics Research Associates, 1979.

Residential Infrastructure: Bacon, 1981.

Transit: Institute for Public Administration, 1980.

Waste Disposal: Albert, Hansen and Wilkinson, 1972; Dales, 1970; DeLucia, 1974; Hanke and Wentworth, 1981.

Water: Angelides and Bardach, 1978; Barry, 1976; Carey, 1976; Grover, 1980; Hanke, 1981 and 1976; Hoggan, 1977; Keller, 1977.

Water: *(see User Fees)*

Grossman, 1979b; Keller, 1977; Kish, 1980; Lake, 1979.

Bibliography

Abt Associates, Inc., *National Rural Community Facilities Assessment Study*, Boston, MA. 1980.

Abouchar, Alan, "A New Approach to the Evaluation and Construction of Highway User Charges." *Eastern Economic Journal*, 1974, pp.34–38.

Adams, Charles F., Jr., and Dan L. Crippen, "The Fiscal Impact of General Revenue Sharing on Local Goverments," unpublished report prepared for the Office of Revenue Sharing, U.S. Department of the Treasury, November, 1978.

Advisory Commission on Intergovernmental Relations, *Significant Features of Fiscal Federalism, 1979–80 Edition*, Washington, DC: GPO, 1980.

_____, *Restructuring Federal Assistance: The Consolidation Approach*, Bulletin No. 79–6, October, 1979.

_____, *The Intergovernmental Grant System: An Assessment and Proposed Policies*, (B-1), September, 1978.

_____, *State Mandating of Local Expenditures*, (A-67), July, 1978.

_____, *A Catalog of Federal Grant-in-Aid Programs to State and Local Governments: Grants Funded FY 1975*, (A-52a), October, 1977.

_____, *Categorical Grants: Their Role and Design*, (A-52), May, 1977.

_____, *Improving Federal Grants Management*, (A-53), February, 1977.

_____, *The Intergovernmental Grant System as Seen by Local, State and Federal Officials*, (A-54), March, 1977.

_____, *Community Development: The Workings of a Federal-Local Block Grant*, (A-57), March, 1977.

_____, *The States and Intergovernmental Aids*, (A-59), February, 1977.

_____, *Federal Grants: Their Effects on State-Local Expenditures, Employment Levels, and Wage Rates*, (A-61), February, 1977.

_____, *Pragmatic Federalism: The Reassignment of Functional Responsibility*, (A-49), July, 1976.

_____, *Local Revenue Diversification: Income Sales Taxes and User Charges*, (A-47), June, 1974.

_____, *Multistate Regionalism*, (A-39), April, 1972.

_____, *Measuring the Fiscal Capacity and Effort of State and Local Areas*, (M-58), March, 1971.

Alaska, Legislative Affairs Agency, *State Assistance to Local Governments on Bonding Problems*, Juneau, AL: Legislative Council, Legislative Affairs Agency, January, 1967.

_____, Legislative Budget and Audit Committee, *Alaska's Public Corporations: A Framework for Assessment*, prepared by the Institute for Public Administration, New York, January, 1982.

Aldrich, Mark, *A History of Public Works in the United States, 1790–1970*, Washington, DC: U.S. Department of Commerce, 1979.

Alexander, Governor Lamar, *The Five Year Capital Budget for the State of Tennessee, 1980–81, 1984–85*, Nashville: The Government of the State of Tennessee, January, 1980.

Amara, Ray, "Strategic Planning in a Changing Corporate Environment," *Long-Range Planning*, Vol. 12, February, 1979, pp. 2–16.

Amdursky, Robert S., *Municipal Bond Law: Basics and Recent Developments: A Course Handbook*, New York: New York Practicing Law Institute, 1981.

American Bankers Association, Bank Management Committee, *A Guide for Developing Municipal Bond Credit Files*, New York, 1968

American Enterprise Institute, *Waterway User Charges*, Washington, DC, 1977.

American Public Works Association, *Revenue Shortfall: The Public Works Challenge of the 1980s*, Chicago, 1981.

_____, *Administration of State Capital Improvement Programs: Nine Selected Profiles*, Chicago, 1979.

_____, A History of Public Works in the United States, Chicago, 1976.

American Society of Planning Officials, *Local Capital Improvements and Development Management: Literature Synthesis*, Washington, DC: GPO, July, 1977, prepared for the U.S Department of Housing and Urban Development and the National Science Foundation.

Angelides, Sotirios, and Eugene Bardach, *Water Banking: How to Stop Wasting Agricultural Water*, San Francisco: Institute for Contemporary Studies, 1978.

Anthony, Robert N., and Regina E. Herzlinger, *Management Control in Non-Profit Organizations*, Richard D. Irwin, Homewood, Illinois, 1980.

Armstrong, J. Scott, *Long Range Forecasting*, John Wiley and Sons, New York, 1978.

Aronson, J. Richard, *Determining Debt's Danger Signals*, International City Management Association, Management Information Service, Vol. 8, no. 2, December, 1976.

Aronson, J. Richard, and James R. Marsden, "Duplicating Moody's Municipal Credit Rating," *Public Finance Quarterly, Vol. 8, no. 1, January, 1980, pp. 97–106.*

Aronson, J. Richard and Eli Schwartz, *Management Policies in Local Government Finance*, Washington, DC: International City Managers Association, 1975.

Artz, Robert M., and Hubert Bermond, "The Fee" in *Guide to New Approaches to Financing Parks and Recreation*, New York, Acropolis, 1970.

Ascher, William, *Forecasting: An Appraisal for Policy Makers and Planners*, Johns Hopkins Press, 1978.

Association of the Bar of the City of New York, Committee on Municipal Affairs, *Local Finance Project: Proposals to Strengthen Local Finance Law in New York State*, New York, November, 1978.

Bacon, Kevin, "Paying for Public Facilities after Proposition Thirteen," *Western City*, August, 1981, pp. 8–11.

Badgley, Robin F., and R. David Smith, *User Charges for Health Services*, Toronto: Ontario Council of Health, 1979.

Baer, Jon A., "Municipal Debt and Tax Limits: Constraint on

Home Rule," *National Civic Review*, Vol. 70, no. 4, April, 1980, pp. 204–10.

Barnes, Garry, "Going Into Lease Financing," *The Bankers Magazine*, Vol. 164, no. 4, July/August, 1981, pp. 9–14.

Barr, James L., "Rational Water Pricing in the Tucson Basin," *Arizona Review*, Vol. 25, October, 1976.

Bates, John C., and Barry M. Wolfson, "New Mortgage Bond Arbitrage Restrictions: The Problems and Potential Solutions," *Weekly Bond Buyer*, September, 1981, pp. 5, 45.

Baumol, William J., "What Economists Can Teach Managers," *American Economic Review*, Vol. 84-3, May, 1976, pp. 431–449.

Beals, Alan, *Cities: Infrastructure Problems and Needs*, Washington, DC: National League of Cities, September 10, 1981.

Benson, Earl D., "Municipal Bond Interest Cost, Issue, Purpose and Proposition 13," *Governmental Finance*, September, 1980, pp. 15–19.

Beyer, Stuart, *Statewide Public Authorities in New York: The Question of Control*, Ph.D. Thesis, State University of New York at Albany, University Microfilms, NY, 1972, (Ann Arbor, MI).

Bologna, Jack, "Why Managers Resist Planning," *Managerial Planning*, January/February 1980, pp. 51–56.

Boyett, Arthur S., and Gary A. Giroux, "The Relevance of Municipal Financial Reporting to Municipal Security Decisions," *Governmental Finance*, May, 1978, pp. 29–34.

Braun, J. Peter, et al., "Dollars for Debt: A Case for Planned Debt Management," *New Jersey Municipalities*, November, 1978, pp. 1–7.

Brealey, Richard, and Stewart Myers, *Principles of Corporate Finance*, McGraw Hill, New York, 1981.

Brody, Susan E., *Federal Aid to Energy Impacted Communities: A Review of Related Programs and Legislative Proposals*, Cambridge, MA: Lab of Architecture and Planning, MIT, 1977.

Brookshire, David S., and Ralph C. D'Arge, "Adjustment Issues of Impacted Communities or, Are Boomtowns Bad?" *Natural Resources Journal*, Vol. 20, July, 1980.

Browne, Lynn E., and Richard F. Syron, "The Municipal Market

Since the New York City Crisis," *New England Economic Review*, July, 1979, pp. 11–26.

Buckley, Michael Patrick, "Assessing the Issues and Trends in Public Utilities Financing: Planning and Policy Considerations for State and Local Governments in Oregon," master's thesis, University of Oregon, Salem, 1982.

Burchell, Robert W., and David Listokin, *Cities Under Stress*, New Brunswick, NJ: Rutgers University, Center for Policy Research, 1981.

Burr, Eugene, *Preparation of a Capital Improvement Program*, MTAS Technical Report, Knoxville: Municipal Technical Advisory Service, Institute for Public Service, University of Tennessee Municipal League, November, 1975.

Buss, Terry F., *Innovative Financial Mechanisms for Urban Economic Development: A Bibliography*, Public Administration Series, Monticello, IL: Vance Bibliographies, May, 1981, Bibliography P-732, p. 14.

———, *Public/Private Partnerships for Urban Economic Development: A Bibliography*, Public Administration Series, Monticello, IL: Vance Bibliographies, May, 1981, Bibliography P-731.

Cagan, Phillip, "The Interest Savings to States and Municipalities from Bank Eligibility to Underwrite All Nonindustrial Municipal Bonds," *Government Finance*, May, 1978, pp. 40–48.

Carey, D.I., "Conservation Water Pricing for Increased Water Supply Benefits," *Water Resources Bulletin*, Vol. 12, December, 1976, pp. 111–123.

Centaur Associates, Inc., *Economic Development Administration Title 1 Public Works Programs Evaluation*, Washington, DC, May, 1979.

Choate, Pat, "Special Report on U.S. Economic Infrastructure," unpublished paper, The House Wednesday Group, May 18, 1982.

———, *As Time Goes By: The Costs and Consequences of Delay*, Columbus, OH: The Academy of Contemporary Problems, 1980.

———, "Urban Revitalization and Industrial Policy: The Next Steps," testimony before the Subcommittee on the City, Committee on Banking, Finance, and Urban Affairs, U.S.

House of Representatives, hearings on Urban Revitalization and Industrial Policy, Washington, DC, September 7, 1980.

Choate, Pat, and Susan Walter, *America in Ruins: Beyond the Public Works Pork Barrel*, Washington, DC: Council of State Planning Agencies, 1981.

Citizen's Budget Commission, Inc., *A Plan to Expedite the Rebuilding of New York City's Infrastructure*, New York, 1979.

Clyde, Larry F., "Statement before the Senate Committee on Finance," U.S Senate, Washington, DC, 1982.

Coltman, Edward, and Shelley Metzenbaum, *Investing in Ourselves: Strategies for Massachusetts: A Report to the Task Force on Public Pension Investments of the Massachusetts Social and Economic Opportunity Council*, Council of State Planning Agencies, Washington, DC, 1979.

Colorado, Final Report of the Governor's Blue Ribbon Panel, *Colorado: Investing in the Future*, July, 1981.

Congressional Budget Office, *The Interstate Highway System: Issues and Options*, Washington, DC: GPO, June, 1982.

———, *Countercyclical Uses of Federal Grant Programs*, Washington, DC: GPO, November, 1978.

Congressional Research Service of the Library of Congress, *Review of Title V Commission Plans*, Washington, DC: GPO, 1977.

CONSAD Research Corporation, *A Study of Public Works Investment in the United States*, Washington, DC: GPO, 1980. In volumes.
  Vol. 1 *Historical Analysis of PWI Trends and Financing Mechanisms.*
  Vol. 2 *Analysis of Maintenance, Condition and Financing of Urban Capital Stock.*
  Vol. 3 *Effects of Federal Capital Grants on the State-Local Functions: Water Systems, Sewer Systems, Streets and Highways, Bridges and Mass Transit.*

Corr, Arthur, "Capital Investment Planning," *Financial Planning*, August, 1980, pp. 12–15.

Council for Urban Development, *Coordinated Urban Economic Development*, Washington, DC, 1978.

Council of State Community Affairs Agencies, *State Financial Management Resource Guide*, Washington, DC, 1980.

Crinder, Robert A., *The Impact of Inflation on State and Local*

Government, Urban and Regional Development Series, no. 5, Columbus, OH: Academy for Contemporary Problems, 1978.

Daily Bond Buyer, New York: Daily Bond Buyer, issued weekdays.

Dales, J.H., *Pollution, Property and Prices*, Toronto: University of Toronto Press, 1970.

Daniels, Belden, and Nancy Barbe, *New England Innovation: Paradigm for Reindustrialization*, Cambridge, MA: Counsel for Community Development, Inc., February 9, 1981.

Darst, David M., *The Handbook of the Bond and Money Markets*, 2d edition, New York: McGraw-Hill, 1981.

DeLucia, R.J., *An Evaluation of Marketable Effluent Permit Systems: Final Report*, Washington, DC: U.S. Environmental Protection Agency, GPO, 1974.

Denver Regional Council of Governments, *Capital Improvements Programming for Local Governments*, Denver, 1975.

Devoy, Robert, and Harold Wise, *The Capital Budget*, Washington, DC: Council of State Planning Agencies, 1979.

Donnelly, Robert M., "Strategic Planning For Better Management," *Managerial Planning*, November/December, 1980, pp. 3–6, 41.

Dossani, Nazir and Wilbur Steger, "Trends in U.S. Public Works Investment: Report on a New Study," *National Tax Journal*, Vol. 33, no. 2, June, 1980, pp. 97–110.

Douglas, Scott, "Determinants of Capital Budgets," paper presented at the 1977 national meeting of the American Society for Public Administration, Atlanta.

Downing, Paul B., *User Charges and Service Fees*, Tallahassee: Florida State University Press, 1980.

Dyl, Edward A., and Michael D. Joehnk, "Leasing As a Municipal Finance Alternative," *Public Administration Review*, November/December, 1978, pp. 557–62.

Economic Development Administration of the U.S. Department of Commerce, *An Updated Evaluation of the EDA-Funded Industrial Parks—1968–74*, Washington, DC: GPO, 1974.

Economics Research Associates, *Impacts of Fees and Changes on Urban Recreation and Cultural Opportunities*, San Francisco, 1979.

Edelman, Seth, *The Career of the New York State Public Authority*, Albany, NY: State University of New York at Albany, 1976.

Edmonds, Charles P., and William P. Hoyd, "Industrial Development Bond Financing," *Financial Executive*, April, 1981.

Executive Office of the President, *Small Community and Rural Development Policy*, Washington, DC: GPO, December 20, 1979.

———, Office of Management and Budget, *Managing Federal Assistance in the 1980s*, Washington, DC: GPO, March, 1980.

———, Office of Management and Budget, *Special Analyses, Budget of the United States Government—Fiscal Year 1980*, Washington, DC: GPO.

———, Office of Management and Budget, *Public Works as Countercyclical Assistance*, Washington, DC: GPO, November, 1979.

———, Office of Management and Budget, *Reorganization Study of Local Development Assistance Programs*, Washington, DC: GPO, December, 1978.

Feldman, Paul, "On the Optimal Use of Airports in Washington, D.C.," *Socio-Economic Planning Science, Vol. 1, 1967, pp. 21–39.*

Feldstein, Martin, "Distributional Equity and the Optional Structure of Public Prices," *American Economic Review*, March, 1972, pp. 231-247.

Feuer, Albert, "Motor Fuel Tax Alternatives," *State Government*, 1978, pp. 11–17.

Finck, John A., and Howard Pike, *Infrastructure Rehabilitation: Where Do We Go from Here?* Albany, NY: New York State Department of Environmental Conservation, October 28, 1981.

Fischer, Philip J., et al., "Risk and Return in the Choice of Revenue Bond Financing," *Governmental Finance*, September, 1980, pp. 9–13.

Forbes, Ronald W., and John E. Peterson, *State Credit Assistance to Local Governments*, Boston: First Boston Corporation, 1978.

———, *Cost of Credit Erosion in the Municipal Bond Market*, Chicago: Municipal Finance Officers Association, December, 1975.

Forbes, Ronald W., and Edward F. Renshaw, *State Bond Banks: Review of Present Developments and Needs for Bond Banks*, Chicago: Municipal Finance Officers Association, 1972.

Forty States Eye Motor Fuel Tax Boost, Washington, DC: Highway Users Federation, January, 1981.

Foster, Robert, *State Responses to the Adverse Impacts of Energy Development in Wyoming*, Report published by Laboratory of Architecture and Planning, MIT, Cambridge, MA, 1977.

Fujardo, Richard P., *Capital Budgeting: Guidelines and Procedures*, Research Reports in Public Policy, no. 9, Santa Barbara: Urban Economics Program, University of California at Santa Barbara, July, 1976.

Galambos, Eva, and Arthur Schreiber, "Pricing for Local Government: User Charges in Place of Taxes" in *Making Sense Out of Dallas*, Washington, DC: National League of Cities, 1978.

Garfield Schwartz Associates Inc., *Local Infrastructure Planning in Maryland*, Report for Maryland Department of State Planning, Baltimore, MD, May, 1982.

Getzels, Judith, and Charles Thurow, *Local Capital Improvements and Development Management: Analyses and Case Studies*, Chicago: American Planning Association, June, 1980.

Gilmore, John S., and Mary K. Duff, *Boom Town Growth Management: A Case Study of Rock Springs–Green River, Wyoming*, Boulder: Westview Press, 1975.

Glaser, Sidney, "New Jersey's Limits on State and Local Spending: A Model for the Nation," *New Jersey Municipalities*, November, 1978.

Godwin, Steven, and George M. Peterson, *Infrastructure Inventory and Condition Assessment: Tools for Improving Capital Planning and Budgeting*, Urban Institute, Washington, DC, 1982.

Gold, Steven D., *How Restrictive Have Limitations on State Taxing and Spending Been?* Legislative Finance Paper no. 13, Denver: National Conference of State Legislatures, January, 1982.

———, *Trends in the Magnitude and Character of State Debt*, Denver: National Conference of State Legislatures, 1981.

_____, *Property Tax Relief*, Lexington, MA: Lexington Books, D.C. Heath, 1979.

Gramlich, Edward, *Benefit-Cost Analysis of Government Projects*, Prentice Hall, Englewood Cliffs, NJ, 1981.

Gramlich, Edward M., and Harvey Galper, "State and Local Fiscal Behavior and Federal Grant Policy" in *Brookings Paper on Economic Activity, no. 1*, 1973.

Grossman, David A., *The Future of New York City's Capital Plan*, Washington, DC: The Urban Institute, 1979a.

_____, *Water Resources Priorities for the Northeast*, Washington, DC: The Consortium of Northeast Organizations, September, 1979b.

Grover, Kathryn, "Wholesale Water Pricing: A Cost-to-Serve Plan That Works," *American City and County*, Washington, DC: November, 1980.

Groves, Sanford, *Evaluating Local Government Financial Condition: Handbook Z, Financial Trend Monitoring System*, International City Managers Association, Washington, DC, 1980.

Hall, William K., "Changing Perspectives on the Capital Investment Process," *Long-Range Planning*, Vol. 12, February, 1979, pp. 37–40.

Hanke, S.H., "On the Current Crisis in Urban Water Supply," unpublished paper, Washington, DC: President's Council of Economic Advisors, 1981.

_____, *Options for Financing Water Development Projects*, Baltimore: Johns Hopkins University Press, 1976.

Hanke, S.H., and R.W. Wentworth, "On the Marginal Cost of Wastewater Services," *Land Economics*, November, 1981, pp. 196–210.

Harrington, John, *Packaging Housing Mortgage Loans: Strategies for California*, Washington, DC: Council of State Planning Agencies, 1979.

Hatry, Harry P., *Local Government Capital Infrastructure Planning: Current State-of-the-Art and State-of-Practice*, Washington, DC: The Urban Institute, 1980.

_____, *Maintaining the Existing Infrastructure*, Washington, DC: The Urban Institute, August 28, 1980.

Hendershott, Patric H., and Timothy W. Koch, *An Empirical*

Analysis of the Market for Tax-Exempt Securities, New York: New York University, Graduate School of Business Administration, Center for the Study of Financial Institutions, 1977.

Henion, Lloyd, and Mark Ford, "Financing Highway Maintenance," *Journal of Contemporary Studies,* Vol. 4, no. 2, Spring, 1981, pp. 38–47.

Henriques, Diana, "Danger: Public Authority at Work," *New Jersey Reporter,* February, 1982, pp. 6–28.

Herships, David, and Leon Karvelis, *Effects of the Reagan Administration's Economic Recovery Plan on the Credit Standing of State and Local Governments,* New York: Merrill Lynch, Pierce, Fenner and Smith, Inc., December, 1981.

Higgins, James M., "Strategic Decision Making: An Organizational Behavioral Perspective," *Managerial Planning,* March/April, 1978, pp. 9–13.

Higgins, Tom, "Road Pricing: Should and Might It Happen?" *Transportation,* Vol. 8, June, 1979, pp. 1–17.

Hoggan, D.H., *A Study of Feasibility of State Water Use Fees for Financing Water Development,* Logan, UT: College of Engineering, Utah State University, 1977.

Holland, Stuart, *The State as Entrepreneur,* London: Weidenfeld and Nicholson, 1972.

Holloway, Clark, and William King, "Evaluating Alternative Approaches to Strategic Planning," *Long-Range Planning,* Vol. 12, August, 1977, pp. 74–78.

Howard, Kenneth S., *Changing State Budgeting,* Lexington, KY: Council of State Governments, 1973.

Howell, James M., and Charles F. Stamm, *Urban Fiscal Stress: A Comparative Analysis of 66 U.S. Cities,* Lexington, MA: Lexington Books, D.C. Heath, 1979.

Hoyle, Robert S., "Capital Budgeting Models and Planning: An Evolutionary Process," *Managerial Planning,* November/December, 1978, pp. 78–89.

Hubbell, Kenneth L., editor, *Fiscal Crisis in American Cities: The Federal Response,* Cambridge, MA: Ballinger Publishing Company, 1979.

Humphrey, Nancy, George E. Peterson and Peter Wilson, *The Future of Cincinnati Capital Plant, America's Urban Capital*

Stock Series, vol. 3, Washington, DC: The Urban Institute, 1979.

Humphrey, Nancy, and Peter Wilson, "Capital Stock Condition in Twenty-Eight Cities," Washington, DC: The Urban Institute, unpublished, February, 1980.

Illinois Bond Watcher, Springfield, IL: Illinois Economic and Fiscal Commission, July, 1981.

Illinois General Assembly, Joint Committee on Long-Term Debt, *Report,* Springfield, IL, January 10, 1979.

Ingram, Robert, and Ron Copeland, "State Mandated Accounting, Auditing and Finance Practices and Municipal Bond Ratings," *Public Budgeting and Finance,* Spring, 1982, pp. 21–33.

Institute of Public Administration, *Financing Transit: Alternatives for Local Government: Executive Summary,* Washington, DC: GPO, 1980.

Irwin, David T., "Debt Management for State Government," *State Government,* 1979, pp. 9–17.

Jarrett, James E., and Jimmy E. Hicks, *The Bond Bank Innovation: Maine's Experience,* Lexington, KY: Council of State Governments, February, 1977.

Jones, Benjamin, *Restoring Municipal Credit: The New Jersey Qualified Loan Bond Program,* Lexington, KY: Council of State Governments, June, 1978.

Katzman, Martin T., "Measuring the Savings From Municipal Bond Banking," *Governmental Finance,* March, 1980, pp. 19–25.

———, "Municipal Bond Banking: The Diffusion of a Public Finance Innovation," *National Tax Journal,* Vol. 33, no. 2, June, 1980, pp. 149–60.

Kaufman, Henry, *The Crowding of the Municipal Bond Market,* Salomon Brothers, New York City, August, 1981.

Kaus, Robert, "Jobs for Everyone," *Harpers,* October, 1982, pp. 11–17.

Kaynor, Edward R., "Uncertainty in Water Resources Planning in the Connecticut River Basin," *Journal of the American Water Resources Association,* December, 1978.

Keller, Charles W., "Pricing of Water," *Journal of the American Water Works Association,* Vol. 69, January, 1977, pp. 92–103.

Kidwell, David S., and Patric H. Hendershott, "The Impact of Advanced Refunding Bond Issues on State and Local Borrowing Costs," *National Tax Journal*, Vol. 3, no. 1, March, 1978, pp. 93–100.

Kieschnick, Michael, *Taxes and Growth*, Washington, DC: Council of State Planning Agencies, 1981.

Kimball, Ralph C., "The Effect of a Taxable Bond Option on Borrowing Costs of State and Local Governments in the Northeast," *New England Economic Review*, March, 1978, pp. 21–31.

_____, "Commercial Banks, Tax Avoidance, and the Market for State and Local Debt Since 1970," *New England Economic Review*, January/February, 1977a, pp. 21–32.

_____, *Commercial Bank Demand and Municipal Bond Yields*, Boston: Federal Reserve Bank of Boston, 1977b.

Kirkland, Kenneth J., *Contracting Out Local Government Services: Policy Considerations for State Legislatures*, National Conference of State Legislatures, Denver, CO, 1982.

Kish, T., "A Look at Self-Supporting Utilities," *Water Pollution Control Federation Journal*, Vol. 52, no. 11, November, 1980.

Klapper, Byron, "Municipal Commercial Paper," *Government Finance*, September, 1980, pp. 10–15.

Kolb, Klaus J., "Economic Development in Alaska: Responsibilities for Providing Infrastructure," unpublished paper, Kennedy School of Government, Harvard University, May, 1982.

Kopcke, Richard, and Ralph C. Kimball, "Investment Incentives for State and Local Governments," *New England Economic Review*, January/February, 1979, pp. 20–40.

Korbitz, William E., editor, *Urban Public Works Administration*, Washington, DC: International City Management Association, 1976.

Kunde, James E., and Daniel E. Berry, "Restructuring Local Economies through Negotiated Investment Strategies," forthcoming *Policy Studies Journal* symposium on "Public Policy for Communities in Economic Crisis."

Lake, Elizabeth R., et al., *Who Pays for Clean Water? The Distribution of Water Pollution Control Costs*, Boulder: Westview Press, 1979.

Lamb, Robert B., and Stephen P. Rappaport, *Municipal Bonds: A*

Comprehensive Review of Tax-Exempt Securities and Public Finance, New York: McGraw-Hill, 1980.

Langton, John, *Toll Road Financing: Description and Policy Implications,* Washington, DC: Association of American Railroads, 1981.

Lehan, Edward Anthony, "The Case for Directly Marketed Small Denomination Bonds," *Governmental Finance,* Vol. 5, no. 9, September, 1980, pp. 3–7.

Leistritz, Larry F., and Steven H. Murdock, *The Socioeconomic Impact of Resource Development: Methods for Assessment,* Boulder: Westview Press, 1981.

Levatino-Donoghue, Adrienne, "Local Bonds for Housing," *Journal of Housing,* Vol. 36, no. 6, June, 1979, pp. 306–9.

Levine, Charles H., and Ira Rubin, editors, *Fiscal Stress and Public Policy,* Sage Publications, Beverly Hills: 1980.

Lindsay, Robert, editor, *The Nation's Capital Needs: Three Studies,* New York: New York Committee on Economic Development, 1979.

Littlefield, S.C., and G.F. Thomsen, "Aircraft Landing Fees: A Game Theory Approach," *The Bell Journal of Economics,* Vol. 8, 1977, pp. 201–232.

Litvak, Lawrence, and Belden Daniels, *Innovations in Development Finance,* Washington, DC: Council of State Planning Agencies, 1979.

Lorange, Peter, and Richard Vancil, "How to Design a Strategic Planning System," *Harvard Business Review,* Sept-Oct 1982, pp. 75–81.

Lu, Catherine, *State Responses to the Adverse Impacts of Energy Development in North Dakota,* Cambridge, MA: Laboratory of Architecture and Planning, MIT, 1977.

Lubick, Donald C., and Harvey Galper, "The Defects of Safe Harbor Leasing and What to do about Them," *Tax Notes,* March 15, 1982.

MacDonald, Keith, editor, *Northeast Urban Infrastructures: A Reader,* Boston: Coalition of Northeast Municipalities, 1982.

McWatters, Ann Robertson, *Financial Capital Formation for Local Governments,* Research Report, 79-3, Berkeley: University of California, March, 1979, Institute of Governmental Studies.

Management Policies in Local Government Finance, Washington, DC: International City Management Association, 1981.

Maryland, Capital Debt Affordability Committee, *Report on Recommended Debt Authorization for Fiscal Year 1983, Submitted to the Governor and the General Assembly of Maryland,* Annapolis, August 1, 1981.

Matson, Morris C., "Capital Budgeting: Fiscal and Physical Planning," *Government Finance,* August, 1976, pp. 42–50, 58.

Matz, Deborah, *Trends in the Fiscal Condition of Cities: 1978–80,* Washington, DC: GPO, 1980.

Mentz, J. Robert, et al., "Leveraged Leasing and Tax-Exempt Financing of Major U.S. Projects," *Taxes,* August, 1980, pp. 553–60.

Mick, Susan R., *User Charges and Fees,* Washington, DC: The Urban Institute, 1981.

Miralia, Lauren M., "Municipal Bond Insurance Gaining in Acceptance," *ABA Banking Journal,* February, 1980, pp. 63, 65, 66.

Mitchell, William E., "Debt Refunding: The State and Local Government Sector," *Public Finance Quarterly,* Vol. 7, no. 3, July, 1979, pp. 323–37.

Moak, Lennox L., *Administration of Local Government Debt,* Chicago: Municipal Finance Officers Association of the United States and Canada, 1970.

Moak, Lennox L., and Kathryn W. Killian, *A Manual of Suggested Practice for the Preparation and Adoption of Capital Budgets by Local Governments,* Chicago: Municipal Finance Officers Association of the United States and Canada, 1964.

Monaco, Lynne, *State Responses to the Adverse Impacts of Energy Development in Colorado,* Cambridge, MA: Laboratory of Architecture and Planning, MIT, 1977.

Moody's Bond Survey, New York: Moody's Investors Service, Inc. weekly. *Moody's Municipal and Government Bonds: News Report,* New York: Moody's Investors Service, Inc. Issued every Tuesday and Friday.

Moody's Municipal and Government Manual—American and Foreign, New York: Moody's Investors Service, Inc. Since 1955, annually.

Morgan Guaranty Survey, *Fiscal Stress for States and Localities*, New York: Morgan Guaranty Trust Co., November, 1981.

Mumy, Gene E., "Issue: Costs and Competition in the Tax-Exempt Bond Market," *National Tax Journal*, Vol. 3, no. 1, March, 1978, pp. 81–91.

Municipal Finance Officers Association of the United States and Canada, *A Guidebook to Improved Financial Management in Smaller Municipalities*, Chicago, August, 1978.

———, *Costs Involved in Marketing State/Local Bonds*, Chicago, 1976.

Mushkin, Selma J., "The Case for User Fees," *Taxes and Spending*, April, 1979, pp. 49–62.

———, editor, *Public Prices for Public Products*, Washington, DC: The Urban Institute, 1972.

———, "Prices as an Alternative to Reorganization," paper presented at Conference on Government Reorganization. Woodrow Wilson School, September, 1977.

Mushkin, Selma J., and Charles L. Vehorn, "User Fees and Charges," *Governmental Finance*, Vol. 6, November, 1977, pp. 61–75.

Mussa, Michael L., and Roger C. Kormendi, *The Taxation of Municipal Bonds: An Economic Appraisal*, Washington, DC: American Enterprise Institute for Public Policy Research, 1979.

National Association of Counties, *Bridging the Revenue Gap*, Washington, DC, 1980.

National Conference of State Legislatures, *Guidelines for Single-Family Tax-Exempt Mortgage Revenue Bonds*, Denver, 1980, p. 30.

National Governors' Association, *Federal Roadblocks to Efficient State Government*, Washington, DC, 1977.

National Governors' Association for Policy Research, *Bypassing the States: Wrong Turn on Urban Aid*, Washington, DC, November, 1979.

National League of Cities, "Capital Budgeting and Infrastructure in American Cities: An Initial Assessment," Washington, DC, 1983.

Naylor, Thomas H., "Organizing for Strategic Planning," *Managerial Planning*, July/August, 1979, pp. 3–9, 17.

Naylor, Thomas H. and Kristin Neva, "The Design of a Strategic Planning Process," *Managerial Planning*, January/February, 1980, pp. 3–7.

_____, "The Planning Audit," *Managerial Planning*, September/October, 1979, pp. 31–37.

Neuner, Edward, Dean Dopp, and Fred Sebold, "User Charges versus Taxation as a Means of Funding a Water Supply System," *Journal of the American Water Works Association*, Vol. 69, 1977, pp. 256–281.

Newsweek, "The Decaying of America," August 2, 1982, p. 25.

Northeast-Midwest Institute, *Urban Water Supply and Sewer Needs in the Midwest*, Washington, DC, December, 1980.

Oates, Wallace E., "The Use of Local Zoning Ordinances to Regulate Population Flows and the Quality of Local Services" in *Essays in Labor Market Analysis*, edited by Orly Ashenfelter and Wallace E. Oates, New York: John Wiley and Sons, 1977.

O'Day, D. Kelly, and Lance A. Neumann, *Assessing Infrastructure Need: The State of the Art*, National Academy of Sciences, Washington, DC, February, 1983.

O'Hare, Michael, "Not on My Block You Don't—Facility Siting and the Strategic Importance of Compensation," *Public Policy*, Vol. 25, no. 4, 1977, pp. 78–93.

O'Hare, Michael and Debra Sanderson, "Fair Compensation and the Boomtown Problem," *Urban Law Annual*, Vol. 14, 1978, pp. 29–37.

Osteryoung, Jerome S., "State General Obligation Bond Credit Ratings," *Growth and Change*, Vol. 9, no. 3, July, 1978, pp. 95–103.

Pagano, Michael, and Richard J. Moore, "Emerging Issues in Financing Basic Infrastructure," unpublished paper, September, 1981.

Pascal, Anthony, *User Charges, Contracting Out, and Privitization in an Era of Fiscal Retrenchment*, P-6471, Santa Monica, CA: The Rand Corporation, April, 1980.

Peterson, George E., *An Examination of State and Local Governments' Capital Demand, Alternative Means of Financ-*

ing 'Public' Capital Outlays and the Impact on Tax-Exempt Credit Markets, Washington, DC: The Urban Institute, n.d., pp. 65–69.

———, Tax-Exempt Financing of Housing Investment, Washington, DC: The Urban Institute, 1979.

Peterson, George E., and Mary John Miller, Financing Infrastructure Renewal: Policy Options, Washington, DC, Urban Consortium, December, 1981.

Peterson, John E., "Has the Municipal Bond Market Undergone Fundamental Change?" unpublished paper, May, 1982.

———, "Current Research in State and Local Government Debt Policy and Management, parts 1, 2, Government Finance, March/June, 1979a, pp. 45–48; November, 1978, pp. 33–35.

———, State Roles in Local Government Financial Management: A Comparative Analysis, Washington, DC: Government Finance Research Center, 1979b.

———, State and Local Government Finance and Financial Management: Compendium of Current Research, Chicago: Municipal Finance Officers Association of the United States and Canada, 1978, p. 27.

———, Watching and Counting: a Survey of State Assistance to and Supervision of Local Debt and Financial Administration, Denver, National Conference of State Legislatures and the Municipal Finance Officers Association, 1977.

———, "Financial Planning for State Government," State Planning Series No. 12, Council of State Planning Agencies, Washington, DC, 1977.

———, Changing Conditions in the Market for State and Local Government Debt, Washington, DC: GPO, 1976.

———, The Rating Game, New York City, Twentieth Century Fund, 1974.

"Preserving U.S. Roads: A Rough Time Ahead," Journal of American Insurance, Winter, 1977–78, pp. 16–20.

Public Securities Association, Fundamentals of Municipal Bonds, New York, 1981.

"Reform of the Municipal Bond Market: Alternatives to Tax-Exempt Financing," Columbia Journal of Law and Social Problems, Vol. 15, no. 3, Fall, 1979, pp. 233–75.

Reilly, James F., *Municipal Credit Evaluation and Bond Ratings Diagnosis, Prognosis and Prescription for Change*, Berkeley, CA: Institute for Local Government, 1967.

Robinson, Donald J., et al., *Municipal Bonds 1981—A Course Handbook*, New York: Practicing Law Institute, 1981.

Rocha, Luis M., "Post-Development Costs in Rural Communities in Alaska," unpublished paper, Kennedy School of Government, Harvard University, May, 1982.

Rothschild, L.F., Unterberg, Towbin, Municipal Research Department, *The Fifty States: Will Budget Problems Continue?* New York, January 11, 1982.

Rumowicz, Madelyn, "In New Jersey: Capital Budgeting and Planning Process," *State Government*, Spring, 1980, pp. 99–102.

Saffran, James S., "Proposition 13: Effect Upon the Bond Market," *Western City*, February, 1979.

Sanderson, Debra, *State Responses to the Adverse Impacts of Energy Development in Texas*, Cambridge, MA: Laboratory of Architecture and Planning, MIT, 1977.

Savas, E.S., "Alternative Institutional Models for the Delivery of Public Services," *Public Budgeting and Finance*, Winter, 1981, Vol. 1. no. 4, pp. 12–20.

Schellenbach, Peter W., and James S. Weber, "Leasing: An Alternative Approach to Providing Governmental Services and Facilities," *Government Finance*, November, 1978, pp. 23–27.

Schilling, Paul R., "Wisconsin Municipal Debt Finance: An Outlook for the Eighties," *Marquette Law Review*, Vol. 63, no. 4, Summer, 1980, p. 539–92.

Schmidt, Richard, "Strategic Planning: Off-Limits for Financial Managers?" *Management Review*, June, 1979, pp. 71–77.

Schneider, Mark, and David Swinton, "Policy Analysis in State and Local Government," *Public Administration Review*, January/February, 1979, pp. 12–17.

Schnell, John F., and Richard S. Krannich, *Social and Economic Impacts of Energy Development Projects: A Working Bibliography*, Monticello, IL: Council on Planning Librarians, 1977.

Schramm, Gunter, *The Value of Time in Environmental Decision Processes*, Ann Arbor, MI: The University of Michigan, November, 1979.

Schulman, Martha A., "Alternative Approaches for Delivering Public Services," *Urban Data Service*, Vol. 14, No. 10, International City Managers' Association, October, 1982.

Schwartz, Gail Garfield, and Pat Choate, *Being Number One: Rebuilding the U.S. Economy*, Lexington, MA: Lexington Books, D.C. Heath, 1980.

Schwertz, Eddie L., Jr., *The Local Growth Management Guidebook*, Washington, DC: The Southern Growth Policies Board, 1979.

Shaul, Marnie S., "The Determinants of City Borrowing," Ph.D. dissertation, Ohio State University, 1980, pp. 135–37.

———, "Capital Financing Options for Local Government," unpublished paper, National Urban Policy Roundtable, 1980.

Sheeran, Burke F., *Management Essentials for Public Works Administrators*, Chicago: American Public Works Association, 1976.

Shepard, Kevin, and Haynes C. Goddard, "New Approaches to Capital Planning and Financing," National Urban Policy Roundtable discussion paper, n.d.

Shoup, Donald C., "Financing Public Investment by Deferred Special Assessment," *National Tax Journal*, Vol. 33, no. 4, December, 1980, pp. 413–29.

Shubnell, Larry, and Bill Cobb, "Creative Capital Financing: A Primer for State and Local Governments," *Resources in Review*, May, 1982, pp. 7–11.

Silber, William L., *Municipal Revenue Bond Costs and Bank Underwriting: A Survey of the Evidence*, New York: New York University, Graduate School of Business Administration, Salomon Brothers Center for the Study of Financial Institutions, 1980.

Small Cities Financial Management Project, *A Debt Management Handbook for Small Cities and Other Governmental Units*, Chicago: Municipal Finance Officers Association of the United States and Canada, 1978.

Smith, Fred L., *Alternatives to Motor Fuel Taxation—Weight-Mileage Taxes*, Washington, DC: Association of American Railroads, 1980.

Smith, Wade S., *The Appraisal of Municipal Credit Risk*, New York: Moody's Investor Service, Inc., 1979.

Solano, Paul, and Steven Hoffman, "Municipal Bond Banking: A Comment," *National Tax Journal*, March, 1982, pp. 64–79.

Stamm, Charles F., and James M. Howell, "Urban Fiscal Problems: A Comparative Analysis of 66 U.S. Cities," *Taxing and Spending*, Fall, 1980, pp. 41–58.

Stanfield, Rochelle, "Building Streets and Sewers Is Easy—It's Keeping Them Up That's the Trick," *National Journal*, May 24, 1980, pp. 1141–1145.

Stanley, David T., *Cities in Trouble*, Columbus, OH, Academy for Contemporary Problems, 1976.

Steiss, Walter Alan, *Local Government Finance: Capital Facilities Planning and Debt Administration*, Lexington, MA: D.C. Heath, 1975.

Sternlieb, George, *Housing Development and Municipal Costs*, New Brunswick, NJ: The Rutgers Center for Urban Policy Research, 1974.

Susskind, Lawrence, and Michael O'Hare, *Managing the Social and Economic Impacts of Energy Development*, Cambridge, MA: Laboratory of Architecture and Planning, MIT, 1977.

———, Task Force on Municipal Bond Credit Rating, *The Rating Game*, New York: Twentieth Century Fund, 1974.

U.S. Bureau of the Census, *1977 Census of Governments*, Washington, DC: GPO, 1980. In 7 volumes.

———, *Statistical Abstract of the United States: 1980*, Washington, DC: GPO, 1979.

———, *State Government Finances*, Washington, DC: GPO, 1964.

———, *Local Government Finances in Selected Metropolitan Areas and Large Countries*, Washington, DC: GPO, 1964.

U.S. Conference of Mayors, *Transit Financing: An Overview of the National Transit Financing Picture in Terms of Federal and State Funding Levels, Fare Structures and Local Revenue Sources*, Washington, DC: GPO, October, 1980 (DOT-P-30-80-34).

U.S. Congress, House Subcommittee on the City of the Committee on Banking, Finance, and Urban Affairs, *City Need and* the *Responsiveness of Federal Grant Programs*, Peggy L. Cuciti Subcommittee Print, Washington, DC: GPO, 1978.

_____, Committee on Public Works, *A National Public Works Investment Policy*, Washington, DC: GPO, December, 1974.

_____, Joint Economic Committee, *Chaos in the Municipal Bond Market*, Washington, DC: GPO, September 28, 1981.

_____, Joint Economic Committee, *Trends in the Fiscal Conditions of Cities, 1979–1981*, Washington, DC: GPO, May, 1981.

_____, Joint Economic Committee, *Public Works as a Countercyclical Tool*, Washington, DC: GPO, 1980.

_____, Joint Economic Committee, Subcommittee on Economic Growth and Stabilization, *Deteriorating Infrastructure in Urban and Rural Areas: Hearing, August 30, 1979*, Washington, DC: GPO, 1979.

_____, Senate Committee on Public Works, United States Senate, 93rd Congress, 2d session, *Construction Delays and Unemployment*, Washington, DC: GPO, 1974.

_____, Senate Committee on Environment and Public Works, *Hearings on the Inland Energy Development Impact Assistance Act of 1977 (S 1493)*, Washington, DC: GPO, 2 parts, August 2, 27, 1977, and May 10, June 19, 1978.

_____, Senate Committee on Governmental Affairs, Subcommittee on Intergovernmental Relations, *Intergovernmental Fiscal Impact of Mortgage Revenue Bonds: Hearing, July 18, 1978*, Washington, DC: GPO, 1979.

U.S. Department of Agriculture, *Rural Development Progress, January 1977–June 1979*, Washington, DC: GPO, 1979.

_____, *Social and Economic Trends in Rural America*, Washington, DC: GPO, October, 1979.

U.S. Department of Commerce, *Establishment of a National Development Bank*, unpublished paper, National Public Advisory Committee on Regional Economic Development, Washington, DC: GPO.

_____, *Governmental Finances in 1979–80*, Washington, DC: GPO, 1981.

_____, *A Study of Public Works Investment in the United States*, Washington, DC: GPO, 1980.

U.S. Department of Defense, *Boom Town Annotated Bibliography*, Washington, DC: The Pentagon, 1981a.

———, *Boom Town Business Opportunities and Management Development*, Washington, DC: The Pentagon, 1981b.

———, Office of Economic Adjustment, *Base Closures: Are the Economic Impact Predictions Realistic? An Analysis of the Post-Closure Economic Impact of Military Installations on Local Communities*, Washington, DC: The Pentagon, 1979.

———, *Communities in Transition, Community Response to Reduced Defense Activity*, Washington, DC: The Pentagon, 1978.

U.S. Department of Housing and Urban Development, *Streamlining Land Use Regulation: What Local Public Officials Should Know*, Washington, DC: GPO, 1980.

———, *The President's National Urban Policy Report*, Washington, DC: GPO, 1980.

———, *Causes and Consequences of Delay in Implementing the Community Development Block Grant Program*, Washington, DC: GPO, June, 1980.

———, *Advance Project Planning for Public Works: A Systematic Approach*, Washington, DC: GPO, 1979.

U.S. Department of Transportation, *The Status of the Nation's Highways: Condition and Performance*, Washington, DC: GPO, January, 1981.

———, Federal Highway Administration, *1981 Federal Highway Legislation: Program and Revenue Options*, Washington, DC: GPO, June, 1980.

———, *Draft Transportation Agenda for the 1980s: The Issues*, Washington, DC: GPO, March, 1980.

———, *Financing Transit: Alternatives for Local Government*, Washington, DC: GPO, July, 1979.

U.S. Environmental Protection Agency, *1990: Preliminary Draft Strategy for Municipal Waste Water Treatment*, Washington, DC: GPO, 1981.

———, *Clean Water: Fact Sheet*, Washington, DC: GPO, April, 1980.

———, *The Cost of Clean Air and Water: A Report to the Congress*, Washington, DC: GPO, August, 1979.

———, *Effective State and Local Capital Budgeting Practices Can Help Arrest the Nation's Deteriorating Infrastructure*, Washington, DC: GPO, November, 1982.

157

, *Better Targeting of Federal Funds Needed to Eliminate Unsafe Bridges,* Washington, DC: GPO, August, 1981.

, *More Can Be Done to Insure that Industrial Parks Create New Jobs,* Washington, DC: GPO, December 2, 1980.

, *Foresighted Planning and Budgeting Needed for Public Buildings Program,* Washington, DC: GPO, September 9, 1980.

, *Perspectives on Intergovernmental Policy and Fiscal Relations,* Washington, DC: GPO, June 28, 1978.

, *Federally Assisted Areawide Planning: Need to Simplify Policies and Practices,* Washington, DC: GPO, March, 1977.

, *Long-Range Analysis Activities in Seven Federal Agencies,* Washington, DC: GPO, December, 1976.

Vance, Mary A., *Municipal Bonds: A Bibliography,* Monticello, IL: Vance Bibliographies, August, 1981.

Vaughan, Roger J., "Federal Tax Policy and State and Local Fiscal Conditions" in *The Urban Impacts of Federal Policies,* Norman Glickman, ed., Baltimore: Johns Hopkins University Press, 1980.

, *Inflation and Unemployment: Surviving the 1980s,* Washington, DC: Council of State Planning Agencies, 1980a.

, "Countercyclical Public Works: A Rational Alternative," testimony given before the Joint Economic Committee of the United States Congress, June 17, 1980b, Washington, DC.

, *State Taxation and Economic Development,* Washington, DC: Council of State Planning Agencies, 1979.

, *Public Works as a Countercyclical Device: A Review of the Issues,* Santa Monica, CA: The Rand Corporation, July, 1976.

Vernez, Georges, and Roger J. Vaughan, *Assessment of Countercyclical Public Works and Public Service Employment Programs,* Santa Monica, CA: The Rand Corporation, 1978.

, and Robert K. Yin, *Federal Activities in Urban Economic Development,* Santa Monica, CA: The Rand Corporation, April, 1979.

, and Burke Burright, and Sinclair Coleman, *Regional Cycles and Employment Effects of Public Works Investments,* Santa Monica, CA: The Rand Corporation, 1977.

Viscount, Francis, *Municipal Bonds: The Need to Regulate*, Washington, DC: National League of Cities, May, 1982.

Vogt, John A., *Capital Improvements Programming: A Handbook for Local Officials*, Chapel Hill: Institute of Government, University of North Carolina, 1977.

Wacht, Richard F., *A New Approach to Capital Budgeting for City and County Governments*, Research Monograph no. 87, Atlanta: College of Business Administration, Georgia State University, 1980.

Wallace, Holly, "Infrastructured: Maintain It Now or Pay the Price Tomorrow" in *City Economic Development*, Washington, DC: National League of Cities, May 12, 1980.

Walter, Susan, ed., *Proceedings of the White House Conference on Strategic Planning*, Council of State Planning Agencies, Washington, DC, 1980.

Walsh, Ann Marie Hauck, *The Public's Business: The Policies and Practices of Government Corporations*, Cambridge, MA: MIT Press, 1978.

Watson, Richard V., *Colorado and New York Evaluate Their Infrastructure Needs and Capital Budgeting Process*, National Conference of State Legislatures, Denver, 1983.

West, Stanley, *Opportunities for Company-Community Cooperation in Mitigating Energy Facility Impacts*, Cambridge, MA: Laboratory of Architecture and Planning, MIT, 1977.

White, Anthony G., *Municipal Bonding and Taxation*, New York: Garland, 1979.

White House Conference on Balanced National Growth, proceedings, Washington, DC: GPO, 1978.

White, Michael J., "Capital Budgeting," in *State and Local Government Finance and Financial Management*, ed. John E. Petersen et al., Chicago: Municipal Finance Officers Association, 1978.

White, Michael J., and Scott Douglas, "An Interpretation of Capital Programming as a Political Process in No-Growth Municipalities," paper presented at the 1975 annual meeting of the American Political Science Association, San Francisco.

White, Sharon S., *Municipal Bond Financing of Solar Energy Facilities*, Washington, DC: GPO, 1980.

Wildavsky, Aaron, *Budgeting: A Comparative Theory of Budgetary Processes*, Boston: Little, Brown and Company, 1975.

Wilson, Peter, *The Future of Dallas' Capital Plant*, Washington, DC: The Urban Institute, 1980.

Wolman, Harold, and Barbara Davis, *Local Government Strategies to Cope with Fiscal Pressure*, Washington, DC: The Urban Institute, 1980.

Wolman, Harold, and George Reigeluth, *Financial Urban Public Transportation: The U.S. and Europe*, New Brunswick, NJ: Transaction Books, 1980.

Worsham, John P., Jr., *Tax-Exempt Mortgage Bonds: Revenue Aids to Homeownership*, Monticello, IL: Vance Bibliographies, November, 1980.

Zeller, Martin, *The Management of Mineral Revenues in the Western Energy-Producing States*, unpublished paper prepared for the Council of State Planning Agencies, Washington, DC, 1982.

Zimmerman, Joseph, *Reassignment of Functional Responsibility, U.S. Advisory Commission on Intergovernmental Relations, Washington, DC: GPO, July, 1976.*

Index

Accountability, of state infrastructure bank, 99–100

Accounting, principles for capital expenditures, 80; techniques, for contracting out, 69

Affirmative action, for public works project, 110

Agriculture, 11, 12, 18, 19

Air pollution, and failure to charge polluters, 55

Airports, expansion of for international trade, 19; part of infrastructure, 1; and public authorities, 107

Alternative capital investments, long-term costs and benefits of, 125

Alternative cost doctrine, 39

Alternative energy sources, 11

Alternative financing, 33, 78, 83, 88

Ambulance services, franchise agreements for, 69

American Public Works Association, study of federal capital funding, 4

American Society of Planning Officials, study by, 81–82

Annual appropriations, 78

Annual budgets, integration with capital plan, 126

Annual operating costs, analysis of for planning, 32

Apparel, United States industry decline due to imports, 18

Aquifers, drilling in and land subsidence, 20

Arable land, 12

Arizona, land subsidence in, 20

As Time Goes By, 5

Assessment of condition, of infrastructure, 84; of existing capital facilities, 126; of local public works, 80

Authorities, *see* Public authorities

Auto emissions, and gasoline prices, 45

Auto imports, 18

Automation, displaces highly paid and low skilled, 12

Average costs, definition of, 41

Baby boom, 16, 53–54

Baltimore, Aquarium in, 62; capital planning and budgeting operations tied together in, 87

Bankers, work at home in future, 15

Bankruptcy, of public authorities, 102, 104

Benefit-cost analysis, definition of, 43

161

Capital expenditures, definition of, 80; and future operating budgets, 126; linked to changes in annual operating budget, 80–83; separate from operating expenditures, 88

Capital facilities, assessment of condition and depreciation rate of, 9; planning for, 1; repair versus new development, 31–32

Capital investment decisions, and overall departmental needs, 79–80

Capital investment plan, all projects included in, 79

Capital investment strategy, purpose of, 10

Capital investments, Federal subsidy for, 72

Capital maintenance needs, assessment of, 83–86; see also Maintenance and repairs

Capital planner, role of, 54

Capital planning, 24–29, 123; and budgeting coordination, 78, 86–88, 126; coordination through state infrastructure bank, 96; definition of, 24–25; document requirements for, 24; establishment of process for, 31–34; goals and objectives of, 31–32; improvement in needed, 5; politics of, 22–34; requirements of, 79–80; saves money, 128; see also Strategic planning; usefulness of, 24–26; what it is, 124

Capital plant, indicators of, condition of, 85

Capital projects, economic factors in planning for, 23;

funds protected for use in, 88; long-term consideration of, 10; status review of, 88; systematic review of, 10

Capitalization rate, see Discount rate

Carbon, replacement for structural steel, 11

Categorical grants, 7, 76, 82; See also Block grants

Ceramics, to produce low-cost construction materials, 12

Charges, assessment of revenues from, 33

Chief executive, see Governor

Choate, Dr. Pat, 5

Cities, and contracting out, 71

City Business Improvement District Law, of New York, 105

City size, and privatization, 71

Clean air, 36, 55, 61

Clean water, sanctions to ensure compliance with regulations on, 100

Clean Water Act Wastewater Construction grants program, 96

Cleveland, capital budget separated from operating budget, 88

Collective good, see Public goods

Colorado, blue-ribbon panel for planning capital spending priorities, 12, 31

Comfort, utility value of, 45

Commerce, U.S. Department of, 82

Community centers, closing of, 76

Community ownership, pride of, 62–63

Competitive bidding, 68

Computer Assisted Design/ Computer Assisted Man-

deter long-term planning, 81

Federal assistance, during recession, 115

Federal capital aid programs, effect on state programs, 81

Federal capital budget, 82

Federal capital grants, compared to state-local construction, 111

Federal categorical programs, *see* Categorical grants

Federal countercyclical public works programs, 108, 109–113, 118, 120

Federal government, 131; depreciation in capital budgeting of, 79; mandates effect on local and state government, 23, 81, 85; policies for public works financing, 3; research and development on maintenance, 86; retrenchment of, 76; stimulates economic activity by expenditure increases, 110; targeting of, 109

Federal grants, curbing of, 1; cuts in and slowed growth of public funds for private use, 59; for highways, 3, 82; influence local choices, 81–82; and planner's role, 124; and national capital budgets, 12; regulations and replacement analysis in, 84; substitute federal for local money, 120

Federal infrastructure bank, 108

Federal infrastructure grants, pooled in state infrastructure bank, 94–100

Federal matching funds, for state stabilization fund, 115–16

Federal programs, financed on year-to-year basis, 26; and

mandated planning, 23, 25; and overlapping responsibilities, 56

Federal stabilization programs, too late, 109–10

Federal to state, delegation of responsibility, 59

Federal wastewater construction grants, 82

Fees, assessment of revenues from, 33

Fellunde (Colorado), survey of future infrastructure demands, 86

Financial factors, affecting viability of city, 85

Financial services, export of United States, 18

Financing mechanisms, constitutional restrictions on, 92; for public works, 92–94

Financing methods, and cost of capital, 83

Financing Public Works in the 1980s, 7–8, 122

Fire safety, public infrastructure category, 2

Fiscal capacity, 131

Fiscal impacts, of proposed projects on future operating expenditures, 88

Fiscal resources available, assessment of, 33

Flood control, not suitable for countercyclical purposes, 120

Food, export of, 19

Forecasting, 52–54, 131; contrasted with planning, 27; of demographic variables for use in planning, 32; of fiscal variables for use in planning, 32; of future level of benefits and costs, 52; limits of, 53–54; of long-term future events, 54; of major eco-

international trade, 19;
maintenance and state stabilization fund, 114
Infrastructure banks, *see*
State infrastructure bank
Inland water transportaion
systems, publicly provided,
60
Insecticides, and endangered
drinking water, 20
Institutional constraints, in
capital project planning, 23;
in state standards, 54, 125
Interagency turf conflicts, and
project planning, 23
Interest rate, and bond market, 49–50
Intergovernmental relations,
132
Intergovernmental revenues,
and financial viability of city,
85
Internal Revenue Service (IRS),
74
International City Management Association (ICMA),
63, 71, 85
International trade, 19; *see
also* Exports, and Imports
Interstate highway program,
3, 83
Intrastate antirecessionary fiscal assistance, part of stabilization fund, 114
Intrastate revenue sharing, to
local government, 7
Investment, costs and payment of, 6; definition of, 47;
ranking of alternatives, 43
Investment-grade bonds, 50
Investment levels, analysis of
for use in planning, 32
Investments in public works,
increase in needed, 122
Investments not made, evaluation of implication of, 79

Investor indemnification,
73–74
Irrigation systems, construction and maintenance of,
105

Japan, work force growth in,
16
Job changing, in future, 17
Job creation, and economic
development projects, 42;
and user tax increase, 108;
and public financing of private sector, 12
Jobs, in import-export in
United States, 18; in information, 12; in manufacturing, 12; and public subsidies
for economic development
projects, 42
Jurisdictions, and public investment projects in own
borders, 93
Justice, public infrastructure
category, 2

Kentucky, highway projects in,
100
Koldene, Ted, 69

Labor force, *see* Work force
Lakes, pollution of, 20
Land-use blueprints, as capi-
tal planning tool, 24–25
Land-use patterns, 10
Lease agreements, risk of, 73
Lease contracts, and state in-
frastructure bank, 94
Lease financing, 132
Lease purchase arrangements,
and constitutional limits on
debt issuance, 74
Leasing, 73, 126
Legal constraints, and ad-
vance of public sector, 59;
and application of objective

standards, 54; on privatization, 63; in state standards, 125
Legal services, contracting out for, 68
Legislative branch, planning responsibility of, 24, 29
Legislative priorities, effect on public agencies, 90
Library services, subsidies for, 70, 76
Life expectancy, increase in, 14
Life-cycle costing, 83, 88
Literate population, external benefit of education, 61
Loans, from state infrastructure bank, 94
Lobbying, in Washington, 76
Lobbyists, 23, 24
Local bond rates, decline in, 111
Local capital facilities, bonds issued by special districts, 105; condition of in city, 85
Local cyclical behavior, 112
Local economic development, 11–12, 110
Local economic downturn, and state infrastructure bank, 128
Local economy, preplanned projects and, 108
Local financing, and privatization, 74; in response to federal grants, 110; from state infrastructure bank, 94–100
Local government, budget problems of, 2; capital budgets of, 78–79; construction funds for, 92; contract monitoring costs of, 75; cost-benefit analyses formalization for, 87; customer of state infrastructure bank, 96; federal mandates effect on, 85; federal tax subsidies for,

71–73; model contracts for contracting out needed by, 68; programs and overlapping responsibilities, 56; ranking of capital choices by, 84; regulatory role in franchising, 69; replacement analysis in, 84–85; retrenchment of, 76; role in rebuilding America, 1, 3; and state infrastructure bank, 97; technical assistance on contracting out, 68; versus public authority, 127
Local growth patterns, and public facilities planning, 32
Local jurisdictions' contributions, to state stabilization fund, 116
Local priorities, in highway program, 81
Local property tax, special assessment added to, 105
Local property values, increased by special assessment districts, 127
Local public capital stock, techniques for monitoring of, 86
Local Public Works Act, 111–12
Local special assessment districts, 9
Long Island, insecticides in drinking water, 20
Long-term capital planning, through state infrastructure bank, 97
Long-term care institutions, for elderly, 18
Long-term economic changes, capital planning for, 24
Long-term planning, 8, and integration with annual budgets, 80
Long-term projects, authoriza-

35–36, 38
Pork-barrel projects, 56, 93
Port Authority of New York and New Jersey, 72, 75
Port facilities, expansion of, 19, 81
Port management, by public authority, 101
Port rehabilitation, public works projects of state stabilization fund, 120
Ports, 1, 107
Post office, discount rate and, 50
Pot-hole filling, contracting out for, 68
Power plants, long-lived assets, 41
Predicting impact of technological changes, 123
Preservation areas, managed to prevent overuse of, 20
President Reagan, 25, 76
President Roosevelt, Works Progress Administration, 1
Price, estimate of minimum value, 45–46
Pricing public facilities, cost-benefit analysis of, 37
Priorities, for public spending, 76
Prioritizing capital choices, 84
Prisons, planning versus managing, 91; privatization of, 62
Private and public control, complex issue, 59
Private construction of service facilities, savings from, 71, 72
Private corporation, plant depreciation of, 79; strategic planning in, 25–26
Private developers, and energy projects in western states, 57, 125–26

Private financing, of public sector, 125
Private firms, cost-sharing with government, 6; role in maintaining public facilities, 6
Private hospitals, and government subsidies cutback, 6
Private investment, and government subsidies, 5–6
Private management of service facilities, savings from, 71, 72
Private markets, and inefficient decisions, 55
Private ownership, of service facilities, 71–73
Private schools, 61
Private sector, contributions to state infrastructure bank, 96; pays construction or maintenance costs, 92; versus public sector, 31
Privatization, 8; agency administering facility, 75; alternative methods of, 63; consequences of, 57; costs and benefits implicit in, 75; definition of, 57; and fiscal prudence, 75; and investment in infrastructure, 74; of public sector, 126; of service delivery, 71–73; standards for, 75; and tax-exempt debt, 75
Procurement, by public authority, 127
Product research and development, costs lowered by, 12
Production facilities, public infrastructure category, 2
Professional planners, role in planning, 23–24, 34
Profit incentives, 36, 70
Program cuts, in ad hoc fashion, 76